Critical Thoughts From a Government Mindset

CHARTRIDGE BOOKS OXFORD

Thoughts With Impact Series

The purpose of the *Thoughts With Impact Series* is to act as a knowledge repository covering the key fields of government practice. The Series also aims to: support the mechanics of knowledge-building in the government sector and public administration; contribute to understanding the building blocks; and identify ways to move forwards.

The Series recognises that it is important that to analyse government practices from a systems theory perspective. We need to view government and public administration from a broader perspective and not depend on the events and circumstances in our own environments. Such broad thinking and analysis should guide and help us focus on outcomes rather than reacting to events as has always been done in the past.

Titles in the *Thoughts With Impact Series* include:

- *Critical Insights From Government Projects (number 1 in the Series)*

- *Critical Thoughts From a Government Perspective (number 2 in the Series)*

- *Critical Insights From a Practitioner Mindset (number 3 in the Series)*

- *Critical Thoughts From a Government Mindset (number 4 in the Series)*

Critical Thoughts From a Government Mindset

DR. ALI M. AL-KHOURI

CHARTRIDGE
BOOKS OXFORD

Chartridge Books Oxford
Hexagon House
Avenue 4
Station Lane
Witney
Oxford OX28 4BN, UK
Tel: +44 (0) 1865 598888
Email: editorial@chartridgebooksoxford.com
Website: www.chartridgebooksoxford.com

Published in 2013 by Chartridge Books Oxford

ISBN print: 978-1-909287-63-1
ISBN digital (pdf): 978-1-909287-64-8
ISBN digital book (epub): 978-1-909287-65-5
ISBN digital book (mobi): 978-1-909287-66-2

Typeset by Domex, India
Printed in the UK and USA

Contents

Author's note

This book was originally published in 2013 by the Emirates Identity Authority, Abu Dhabi, United Arab Emirates. Permission to republish this book is gratefully acknowledged.

The Chapters in this book have previously been published elsewhere:

Chapter 1. Corporate government strategy development: A case study. Copyright © 2012 *Business Management Dynamics*. Al-Khouri, A.M. (2012) 'Corporate Government Strategy Development: A Case Study'. *Business Management Dynamics* (2) 1: 5–24.

Chapter 2. Customer relationship management: Proposed framework from a government perspective. Copyright © 2012 *International Journal of Management and Strategy*. Al-Khouri, A.M. (2012) 'Customer Relationship Management: A Proposed Framework from a Government Perspective'. *International Journal of Management and Strategy* 3 (4): 34–54.

Chapter 3. e-Government strategies: The case of the United Arab Emirates. Copyright © 2012 *European Journal of e-Practice*. Al-Khouri, A.M. (2012) 'e-Government Strategies: The Case of the United Arab Emirates'. *European Journal of e-Practice* 17: 126–150.

Chapter 4. e-Government in Arab countries: A six-stage road map to develop the public sector. Copyright © 2012 *International Journal of Management and Strategy*. Al-Khouri, A.M. (2013) 'e-Government in Arab Countries: A 6-Staged Roadmap to Develop the Public Sector'. *International Journal of Management and Strategy* 4 (1).

Chapter 5. e-Voting in the UAE FNC elections: A case study. Copyright © 2012 *Information and Knowledge Management*. Al-Khouri, A.M. (2012) 'e-Voting in UAE FNC Elections: A Case Study'. *Information and Knowledge Management* 2 (6): 25–84.

Chapter 6. Identity and mobility in a digital world. Copyright © 2013 *Technology and Investment*. Al-Khouri, A.M. (2013) 'Identity and Mobility in a Digital World'. *Technology and Investment* 4 (1).

Chapter 7. Data ownership: Who owns 'my data'? Copyright © 2012 *International Journal of Information Technology*. Al-Khouri, A.M. (2012) 'Data Ownership: Who Owns 'My Data?' *International Journal of Management and Information Technology* 2 (1): 1–8.

Chapter 8. Triggering the smart card reader supply chain. Copyright © 2013 *Technology and Investment*. Al-Khouri, A.M. (2013) 'Triggering the Smart Card Readers Supply Chain'. *Technology and Investment* 4 (2).

Preface

We are living in a constantly changing and evolving world that is full of twists, turns and dynamism. The technological revolution in the past few years has been radical and ground-breaking at all levels. In the middle of all this, governments are faced with multiple and complex challenges to develop efficient and effective systems that address both global and local needs.

Our perspective suggests that in order to develop better government capabilities, we need to make efforts on a global scale to convert the widespread and scattered knowledge from different countries and regions into best practices and standards to develop citizen-centric government models. The field of government practice, however, is still concealed, and access to data in governments is very much restricted. Existing knowledge about government practices is not reported from within. Current publications and literature about government and public administration are limited, unclear, and do not provide sufficient insight into the inner systems and relationships.

This is the specific purpose of the *Thoughts with Impact Series*. Acting as a knowledge repository, the articles in this book explore and document experiences and aim to contribute to the existing body of knowledge in key fields of government practice. We also aim to support the mechanics of knowledge building in the government sector and public administration, and to contribute to understanding the building blocks and identifying ways to get beyond them. Assuming that we can use old-school thinking and depend on what has worked before will definitely lead us to agonising realities!

As a final note, it is important that we analyse government practices from a systems theory perspective. We need to view government and public administration from a broader perspective and not depend on the events and circumstances in our environments. Such broad thinking and analysis should guide and help us focus on outcomes rather than reacting to events as has always been done in the past. Systems theory should always remind us that if we break up an elephant, we won't have a bunch of little elephants!

The articles in this book have been grouped into three categories: (1) strategic management; (2) e-government development and practices; and (3) identity management. In the first category, we present two articles. The first is titled 'Corporate Government Strategy Development', and it provides a detailed case study of a strategy development exercise in a federal government agency that is now considered one of the most successful government organizations in the public sector in the United Arab Emirates. The second article is titled 'Customer Relationship Management', and it proposes an innovative framework to act as a practical management tool and provide a holistic overview of implementation phases, the components of each phase, and the associated critical success factors.

In the second category, we have three articles. The first is titled 'e-Government Strategies', and it provides an overview of the UAE e-Government Strategic Framework 2011–2013 and explains how the UAE government intends to develop a government-owned federated identity management system to support government-to-citizen (G2C) e-government transactions and promote trust and confidence on the Internet. The second article is titled 'e-Government in Arab Countries'. It presents a conceptual six-stage roadmap that illustrates how Arab countries should prioritize their e-government in short and mid-term efforts. The article argues that the proposed road map could support the development of the public sector and the emergence of the Arab bloc as strong, revolutionized, citizen-centric governments. The third article, titled 'e-Voting in UAE FNC Elections', presents a case study of the e-voting system deployment in the Federal National Council elections in the United Arab Emirates (UAE). It provides detailed insights into the phases of the project, from the design phase up to election day.

The third category has three articles. Article One is titled 'Identity and Mobility in a Digital World'. It explores the potential role of government-issued smart identity cards in leveraging and enabling a more trusted mobile communication base. It delves into the identity management infrastructure program in the UAE and how the smart identity card and overall system architecture have been designed to enable trusted and secure transactions for both physical and virtual mobile communications. The second article is titled 'Data Ownership: Who Owns my Data?'. It explores this critical topic and the fact that many personal and non-personal aspects of our day-to-day activities are aggregated and stored as data by both businesses and governments. The aim of the article is to raise awareness and trigger a debate for policy-makers with regard to data ownership and the need to improve existing data protection, privacy laws, and legislation at both national and international levels. The third article, titled 'Triggering the Smart Identity Card Readers' Supply Chain', explores card reader adoption opportunities in both the public and private sectors and attempts to outline the United Arab Emirates government's plans to disseminate card readers and promote their adoption in the government and other industries in the country.

We hope you enjoy reading this book.

Dr. Ali M. Al-Khouri
2013

About the author

Dr. Ali M. Al-Khouri

Dr. Ali M. Al-Khouri heads the Emirates Identity Authority, a UAE federal organisation, as Director General. He received his Engineering Doctorate degree from Warwick University where his research focused on the management of strategic and large-scale projects in the government sector. He is a Certified Project Management Professional and a Chartered Fellow of the British Computer Society. He has been involved in many strategic government development projects, and lately the UAE national ID project as an executive steering board member and the chairman of the technical committee. His main research interests include the application of modern and sophisticated technologies in large contexts, projects management, organisational change and knowledge management.

Part 1
Strategic management

Corporate government strategy development: a case study

Abstract: In this article, we present a case study of one the successful government organisation strategy development exercises in the United Arab Emirates (UAE). The 2010–2013 strategy of Emirates Identity Authority (Emirates ID) supported organisational development and allowed the Emirates ID to become a pioneer in its field of practice. Its strategy was among the reasons behind its selection and winning the Best Federal Authority Award in the 2012 UAE's federal Government Excellence Programme. This article attempts to describe the principles on which the organisational strategy was developed. It also outlines major accomplishments and the strategy's impact on overall organisational performance. We conclude that the new strategy helped the organisation become a successful example in the UAE and that further lessons can be learned from it, as we outline and discuss them to influence the field of practice.

Keywords: *Emirates ID 2010–2013 strategy, strategic planning, strategy development, national identity management systems*

1. Introduction

Strategic planning is a management tool that helps an organisation focus its energy, ensure that members of the organisation are working toward the same goals, and assess and adjust the organisation's direction in response to a changing environment (Bryson, 2011). In short, strategic planning is a disciplined effort to produce fundamental decisions and actions that shape and guide what an organisation is, what it does, and why it does it, with a focus on the future (Bryson, 2011) (see Figure 1.1).

Strategic planning has become prevalent in all parts of the globe. However, practice in the field shows that the strategy development

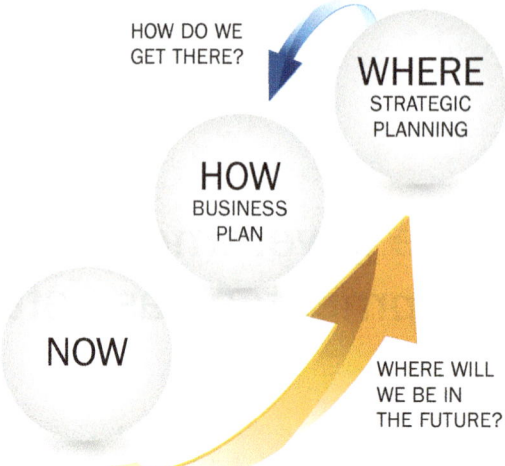

Figure 1.1 Strategic planning focus areas

process itself has become daunting. Ironically, by creating an unfocused strategy, a government can end up wasting money and resources on an effort that will not deliver the desired outcomes. On the other hand, if the government focuses on creating a more focused strategy, subsequent gains should result in energising and moving the organisation toward its mission goals, creating an integrated system by which the strategic plan becomes a reference to measure progress and a system of accountability (Bryson, 2011; Lipman and Lipman, 2006; Steiner, 1997).

Still, to date, many organisations fall into what Ahoy described in 1998 as the 'trap' where most organisational time is spent reacting to unexpected changes instead of anticipating and preparing for them in a carefully crafted strategy. This is referred to in today's terminology as 'crisis management.' Ahoy also referred to the reality that organisations caught off guard may spend a great deal of time and energy playing catch-up games and using up their energy coping with immediate problems, with little energy left to anticipate and prepare for the next challenge. This vicious cycle locks many organisations into a reactive position.

From this perspective, we present a case study of a successful strategy development project in the government field. It provides an overview of how the Emirates Identity Authority (Emirates ID), a federal government organisation in the UAE, developed its globally benchmarked 2010–2013 strategy. This strategy positively affected the overall organisational performance and supported the organisation to become a pioneer in its field of practice. The organisation was awarded the Best Federal Authority in the UAE Federal Government Excellence Programme* in 2012. We also report some of the lessons learned during the first two years of development and execution phases of the strategy. The overall content and discussion in the article can serve as a reference for practitioners in the field.

The article is organised as follows. The next section provides background information to the case study organisation and the strategy context in which it was developed. In Project Achievements, we present the main outcomes and achievements of the new strategy. In Strategy Impact on Emirates ID, we provide an overview of the strategic impact on the overall organisational performance and the key areas that were positivity impacted. In Lessons Learned, we outline some of the key lessons during development and execution of the new strategy. In the Conclusion we summarise the document with some remarks and reflections.

2. The Emirates ID background: the strategic context

The Emirates ID is an integral part of the UAE federal strategy and has a mandate to create a central population database that stores biometric and biographical information for all residents and citizens in the UAE, supplying value-added smart identity cards to all those who enrol in the database. Emirates ID underwent a preliminary strategy development

---

* The UAE Federal Government Excellence Programme (also referred to as Sheikh Khalifa Government Excellence Programme) aims at promoting excellence in the public sector and improving both performance and results based on international criteria. The program uses the European Framework for Quality Management (EFQM) model to support the development of a 'government of excellence' in pursuit of its 2021 vision to become among the best countries in the world (*http://www.skgep.gov.ae/*), (*http://www.vision2021.ae/*).

exercise in early 2007, which resulted in clear goals designed to address the organisation's initial infrastructure, technology and organisational requirements. While this strategy was effective in the start-up phase, achieving the authority's mandate required a more dynamic strategy that would also take into account a required evolution in the operating model.

Therefore, in the fourth quarter of 2009, Emirates ID's leadership team decided to undergo a second strategy development exercise to prepare the authority for the possible challenges that lay ahead (e.g., mass enrolment, infrastructure expansion, e-services facilitation and so on). As a result, the leadership team also decided to trigger the development of the authority's new strategy, taking into account the anticipated challenges while maintaining alignment between its strategy and the UAE federal strategy. This project also enabled Emirates ID to become one of the UAE's first federal entities to successfully implement the customised strategy development framework developed by the UAE Prime Minister's office.

Once the 2010–2013 strategy had been formalised and approved in February 2010, Emirates ID's next challenge was the timely roll-out and effective execution of the strategy. To facilitate this process, Emirates ID formed a Strategy Support Office (SSO) headed by the Director General of the authority and consisting of skilled program managers, and experts and specialists in management and technical areas. The SSO was responsible for:

- Reporting Emirates ID strategy development and execution activities to the Board, the UAE Prime Minister's Offices, and other stakeholders.
- Supervising the cascading of the strategy into departmental operational plans.
- Performing ongoing updates to the strategy.
- Supporting internal communication of the strategy.
- Providing management support to initiative execution.

Over 24 months, the SSO collaborated on various aspects of strategy execution, change management, and stakeholder management, which required involvement in internal initiatives ranging from supporting the development of departmental operating plans to the improvement and monitoring of enrolment and card delivery processes. A number of these initiatives are illustrated in Figure 1.2.

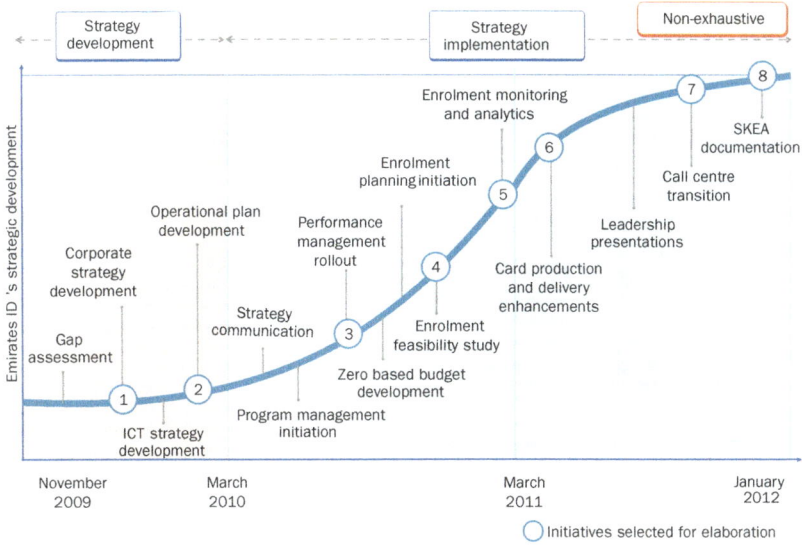

Figure 1.2 Illustration of Emirates ID strategy development and execution journey

3. Project achievements

Project achievements have been segmented into two areas: strategic development and execution, with the former further broken down into key initiatives.

3.1 Corporate strategy development project

Emirates ID's 2010–2013 strategy was developed using the strategy development framework formulated by the UAE Prime Minister's Office and mandated for use by all UAE federal government entities (see Figure 1.3).

The strategy development framework starts with an articulation of the high-level strategic direction embodied by the vision and the mission statements, and the strategic intentions of the federal government. This high-level strategic direction then cascades down through the organisation in a structured manner in the form of strategic objectives, key success factors, initiatives, and activities for departments.

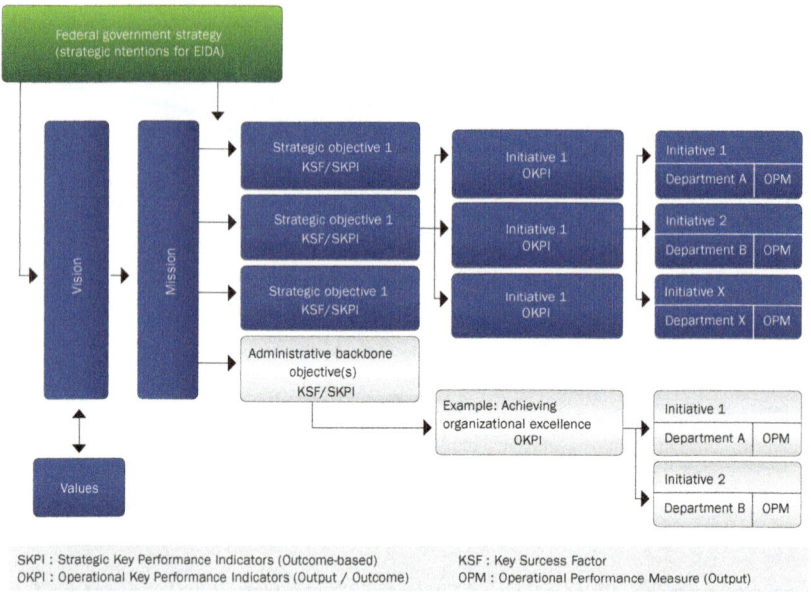

SKPI : Strategic Key Performance Indicators (Outcome-based)   KSF : Key Surcess Factor
OKPI : Operational Key Performance Indicators (Output / Outcome)   OPM : Operational Performance Measure (Output)

Figure 1.3 UAE federal government strategy development framework

The structured framework ensures explicit linkages between every level of the strategy, from the initial idea all the way down to the activities, hence providing a coherent road map to deliver the Emirates ID's mandate. This also provided clarity for staff as to how their work directly contributed to the achievement of a specific element of the strategy.

The leadership team and SSO office applied the above strategy development framework through the project approach to set Emirates ID's new strategic direction. The project approach consists of three distinct phases (see Figure 1.4).

The diagnostic phase started with a thorough review of Emirates ID's existing strategy (2007–2010), as well as discussions involving the

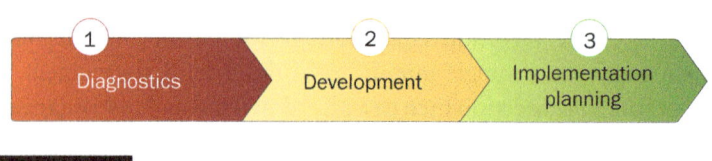

Figure 1.4 Strategy development approach

leadership team to articulate the vision of the organisation. These discussions also focused on how the vision could or should be realised and potential challenges that may be faced along the way. Meanwhile, numerous analyses were conducted to uncover potential internal challenges facing the organisation.

The outcome of these discussions and analyses was an organisational health check that highlighted critical gaps that needed to be taken into account when developing Emirates ID's new strategic direction. Next, the SSO partnered with Emirates ID's leadership team to articulate the current and future operating model, so as to better understand the evolution in the service portfolio and customer base.

A benchmarking exercise was also conducted to enable comparison of Emirates ID's operating model with those of its international peers (see Table 1.1).

Next, the authority embarked on the development phase, which required using output from the diagnostic phase coupled with the strategic intent of the authority to develop the high-level strategic direction for the organisation: mission, vision, and four strategic intentions that were cascaded into four distinct strategic objectives, cumulatively making up the activities of the entire organisation (see Figure 1.5).

Country	E-Gov Rank (2008)[1]	e-ID Card Issuing Entity	Year of First Issue	Benchmarking Appeal
Bahrain	▲	Central informatics organisation	2005	– A regional leader for e-ID and e-Gov – Advanced infrastructure
Belgium	▼	General directorate for institutions and population	2005	– Many integrated functionalities with the card – Very advanced infrastructure
Estonia	▲	Citizens and migration board	2002	– One of the first European countries to issue e-ID cards – Many integrated functionalities with the card – Very advanced infrastructure
Malaysia	▲	National registration department	2001	– First to issue e-ID cards in the world – Many integrated functionalities with the card – Very advanced infrastructure
Singapore	▼	Immigration and checkpoints authority		– Many integrated functionalities with the card – Very advanced infrastructure

Table 1.1 Benchmarking with international peers

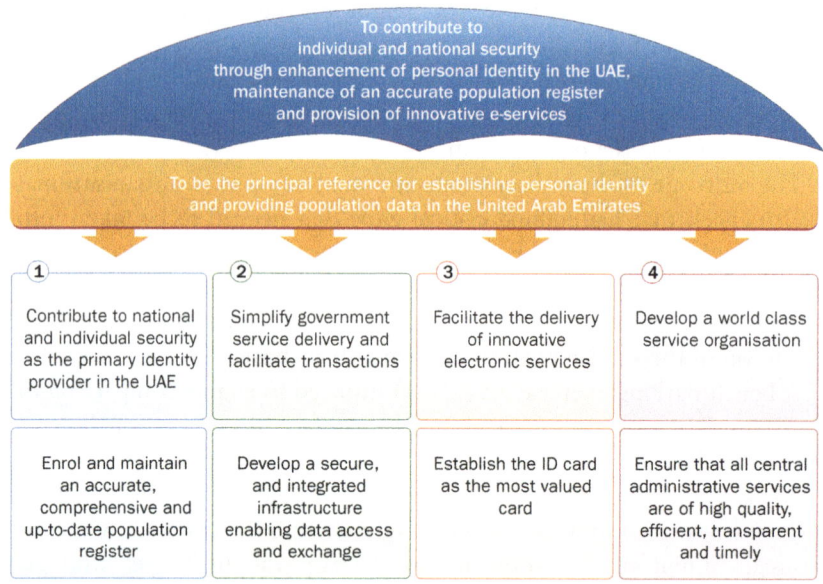

To contribute to
individual and national security
through enhancement of personal identity in the UAE,
maintenance of an accurate population register
and provision of innovative e-services

To be the principal reference for establishing personal identity
and providing population data in the United Arab Emirates

①	②	③	④
Contribute to national and individual security as the primary identity provider in the UAE	Simplify government service delivery and facilitate transactions	Facilitate the delivery of innovative electronic services	Develop a world class service organisation
Enrol and maintain an accurate, comprehensive and up-to-date population register	Develop a secure, and integrated infrastructure enabling data access and exchange	Establish the ID card as the most valued card	Ensure that all central administrative services are of high quality, efficient, transparent and timely

Figure 1.5   Emirates ID's mission, vision, and strategic objectives

For each strategic objective, a series of measurable key success factors (KSFs) – strategic enablers – were also identified to guide progress toward achieving the respective strategic objective to which each belonged. In total, 16 KSFs were established and assigned key performance indicators (KPIs), which would measure the progress in achieving each KSF – hence, the overall strategic objective.

In the implementation planning phase, the team took into account the gaps identified, the shift in operating model requirements, as well as the key success factors for the organisation to define 20 distinct initiatives that would cumulatively address all organisational gaps and operating model requirements, with the aim of ultimately delivering all of the strategic objectives. This structured strategy development process ensured that all proposed initiatives were linked to potential gaps and KSFs, in turn linked to strategic objectives (see Table 1.2).

3.2 Strategy implementation and the Strategy Support Office (SSO)

The SSO played a critical role in the execution, oversight, and close monitoring of numerous initiatives aimed at executing Emirates ID's

Table 1.2 Identification of initiatives for 2010–2013 strategy

Strategic objectives	Enrol and maintain an accurate, comprehensive and up-to-date population register	
Key success factors/SKPI	Gaps	Initiatives/sub initiatives
• Accessibility to enrolment services • Daily enrolment application/ issuance	• Lack of defined plan outlining population forecasts for enrolment area by time.	• Enrolment and renewal planning • Access channel enhancement: establish additional enrolment channels to enables more rapid enrolment. • Staffing ramp up: deploy a clean plan for recruiting, hiring and retaining front and back end staff need to support future enrolment targets. • Core enrolment system upgrades: carry out necessary software and hardware upgrades required to expand daily enrolment capacity. • Enrolment support systems upgrades: update support system required to expand daily enrolment; i.e. on/ offline registration and so on.
• Ability to manage flow • Daily enrolment n/ issuance as a percentage of capacity. • Percentage of target population enrolled.	• Outdated plan for ensuring consistent flow of population into registration centres. • Understaffing in registration centres will prevent ability to enrol expected daily volume in the future. • Software and hardware limitations will constrain capabilities to cross check expected future volumes with MOI database.	
• Accuracy of data • Percentage of registered customers updating status	• Lack of integration mechanisms and automated process to permit continuous updating of the population register; either by government entities or the general population.	• Population Register Accuracy: implementing a plan for managing automated and user driven updates and implement integration layer to allow for secure update/query access.

2010–2013 strategy. In addition to directly supporting the initial execution processes, the SSO also served as advisor to the senior leadership team in managing the teams' communications with key stakeholders, such as the Emirates ID's Board and the UAE Prime

Minister's Office. This section chronologically highlights some of the key initiatives that were supported by the SSO.

After developing the corporate strategy, the next step was to integrate the new strategy with operational activities. This was done via the development of operational plans for each department within the organisation's structure. Such an approach created a link between the day-to-day activities of each department with the initiatives identified in the corporate strategy. Development of the operating plans consisted of the documentation of high-level activities, planned timelines, and milestones, as well as the assignment of individuals accountable for each initiative. Furthermore, KPIs were also defined to measure progress against targeted output and outcomes (see Figure 1.6).

Key benefits of the operational plans were the creation of elaborate work-plans linking the day-to-day activities of departments to the overall strategy, thereby leading to increased departmental accountability for the execution of the corporate strategy, as well as provision of a holistic view on what to expect from various teams over the course of a year. One of the primary challenges was ensuring that department teams adhered to the agreed operational plan commitments and used the operational plans as a means of guiding the departments' daily activities.

Another key initiative led by the SSO was a performance management program which was progressively rolled out over a two-year period to ensure maximum buy-in and adoption within the organisation. A first

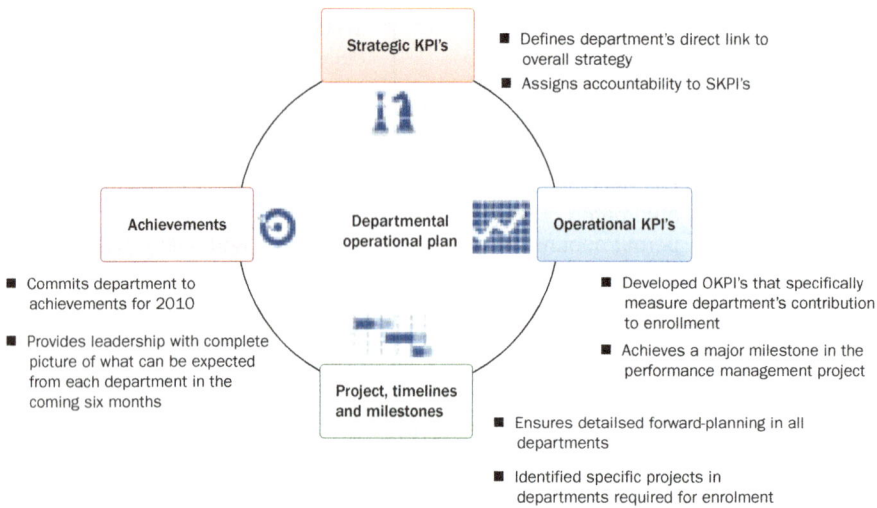

Figure 1.6 Benefits of operational plans

step in the initiative was manual measurement processes for performance indicators identified in the strategy development phase and departmental operating plans. These KPIs also served as the baseline for semi-annual performance reporting to the UAE Prime Minister's Office.

Manual KPI measurement was conducted via customised department-specific Excel spreadsheets that served as a performance tracking and submission tool for each department and also enabled rapid consolidation by the strategy department (see Figure 1.7).

As organisational buy-in increased and organisational needs evolved, the number and scope of performance indicators measured also increased.

Furthermore, an initial subset of performance indicators was identified for automated measurement and was used to develop business requirements that would enable automated measurement of KPIs.

The roll-out of the performance management program has already provided numerous benefits: Not only has it enabled greater transparency around organisational performance and promoted more robust governance, it has also supported compliance with the performance reporting needs of the UAE federal government and created a more results-oriented organisation.

This initiative also posed its fair share of challenges. An initial challenge was obtaining buy-in from departments in the identification of performance indicators and set up of reasonable targets. Furthermore, the departments had to comply with monthly performance reporting timelines and provide supporting documentation, which required close oversight and follow-ups with each department.

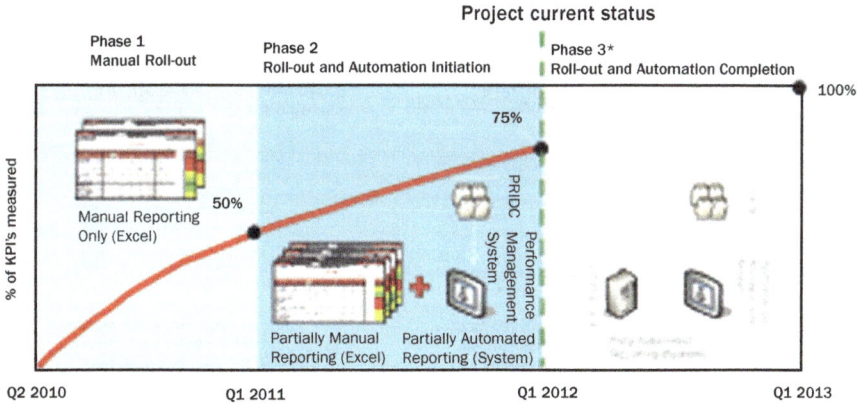

Figure 1.7   Emirates ID performance management road map

One of the pivotal projects carried out during strategy execution was the development of an impact study for the enrolment process re-engineering. As a part of this project, the SSO supported the organisation in gaining stakeholder buy-in and support for the roll-out of a re-engineered enrolment process.

The need for a re-engineered process arose due to challenges in rapidly enrolling citizens and residents into the UAE federal government's 'population register'. Key drivers for these challenges were limited to intake capacity, complex enrolment processes, and lack of robust mechanisms to ensure a regular flow of enrolment applicants into service points.

To remedy this situation, Emirates ID conducted a review of its enrolment processes to identify root causes for bottlenecks, thereafter implementing a process re-engineering to solve these problems (Al-Khouri, 2011). Key benefits of the re-engineered enrolment process could be divided into four key areas: process efficiency, cost optimisation, incremental capacity, and enhanced customer experience (Al-Khouri, 2011). Some key benefits achieved as a part of the re-engineered process are depicted in Figure 1.8 and Table 1.3.

This impact study was instrumental in validating the process re-engineering that had been approved by the Emirates ID's Board in Q3 of

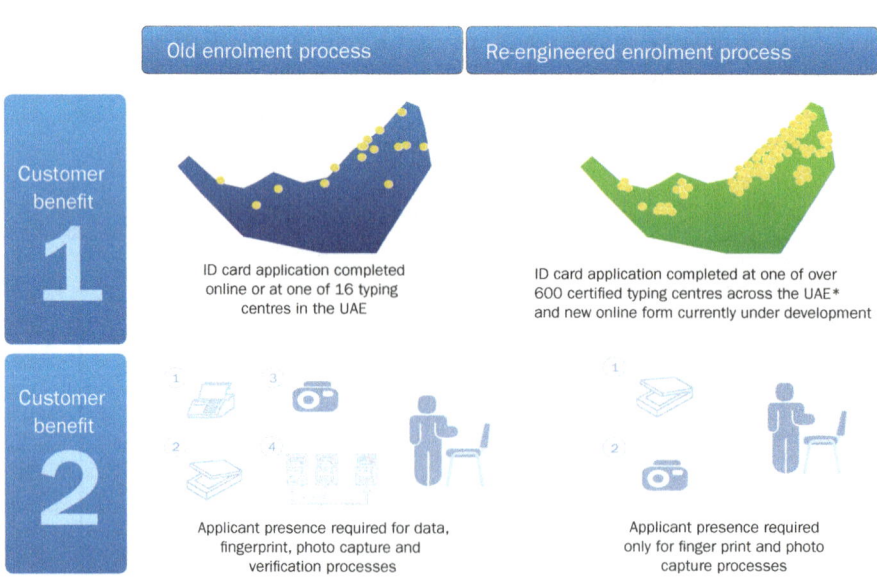

*Planned online application and unified form (currently under development)

Figure 1.8   **Enrolment process optimisation benefits**

Table 1.3	Re-engineering project key benefits

Benefit	Description
Process efficiency.	▪ Intake of ID card applications at over 1,000 certified typing centres. ▪ Limiting of applicant in person presence for biometrics capture only (i.e. photo and fingerprinting), leading to 23 minute reduction in average enrolment time per applicant. ▪ Consequential higher utilisation rate of service point operators and in turn theoretical reduction in average labour cost of 30 AED per application.
Cost optimisation	▪ Forecast savings of over 400 million AED in labour cost over three years via linkage of Emirates ID card to residence visa and key government services (i.e. mandatory issue of 13 million new or renewed ID cards between 2011–2013).
Incremental capacity	▪ Incremental typing centre intake (>2000%). ▪ Fingerprinting and photo capture capacity (133%). ▪ Population Register (PRIDC) processing capabilities (300%). ▪ Card production (100%).
Enhanced customer experience	▪ Automation of process leading to on-line of 'paper trail', enhanced security, and simplified data retrieval. ▪ Availability of appointments allowing customers to better plan service point visits. ▪ Development of online application tracking tool.

2009 and also served as an organisation-wide model for planning and analysis-based solution implementation, which had been one of the key gaps identified in the strategy development phase.

Once approval for the enrolment re-engineering had been achieved and the high-level process re-engineering implemented by the program team, the SSO was called in to help stabilise the new process and ensure the adoption of detailed re-engineered process steps across all parts of the organisation. To help achieve these objectives, the SSO worked hand in hand with the departments and different vendor project teams for more than 15 months to embed new processes and implement mechanisms and tools for monitoring and analysing various stages of the enrolment process. Figure 1.9 outlines the scope of the SSO in strategy management and monitoring activities (see also Table 1.4).

This project served as the backbone of the SSO strategy implementation activities at Emirates ID and had a direct positive effect on its strategic KPI results (e.g., average enrolment capacity per day, average daily

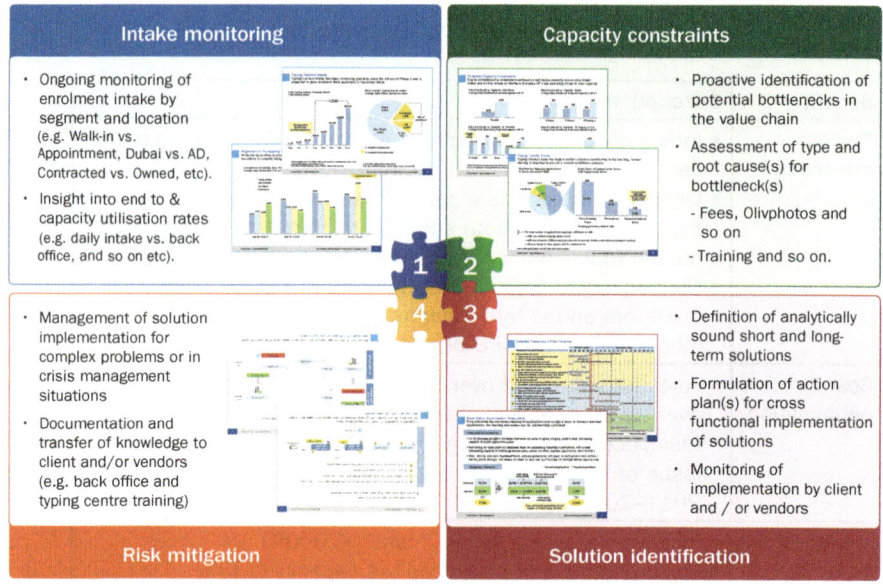

Intake monitoring

- Ongoing monitoring of enrolment intake by segment and location (e.g. Walk-in vs. Appointment, Dubai vs. AD, Contracted vs. Owned, etc).
- Insight into end to & capacity utilisation rates (e.g. daily intake vs. back office, and so on etc).

Capacity constraints

- Proactive identification of potential bottlenecks in the value chain
- Assessment of type and root cause(s) for bottleneck(s)
  - Fees, Olivphotos and so on
  - Training and so on.

Risk mitigation

- Management of solution implementation for complex problems or in crisis management situations
- Documentation and transfer of knowledge to client and/or vendors (e.g. back office and typing centre training)

Solution identification

- Definition of analytically sound short and long-term solutions
- Formulation of action plan(s) for cross functional implementation of solutions
- Monitoring of implementation by client and / or vendors

Figure 1.9   SSO enrolment activities

enrolment, average number of applications processed by the back office and so on). In addition, the SSO's analytical approach served as a valuable insight-sharing mechanism for the bi-weekly management team meetings. Furthermore, preparation of an enrolment management handbook document by the SSO ensured the continued retention of enrolment re-engineering knowledge, should any further enhancements be desired in the future. Figure 1.10 provides a high-level overview of the enrolment process steps.

This initiative was by far the most time consuming and challenging of all the initiatives undertaken in the strategy. A primary driver for the challenging nature of the initiative was a lack of competent project owner(s) accountable for driving the enrolment re-engineering and associated initiatives forward, as well as managing the program via an integrated master plan. This often led to a sense of complacency among both Emirates ID and vendor project teams, necessitating close involvement and even day-to-day management by the SSO. Another issue that heightened the difficulty of the initiative was the quality of external vendor teams and open-ended nature of some vendor contracts. Both of these led to slow execution of even the most basic operational tasks and frequent non-compliance with agreed-upon milestones.

Table 1.4   Strategy activities

No.	Activity description
1	Monitoring of enrolment intake by customer segment and service centre location: • Walk-in customer vs. scheduled customer (appointment) • Enrolment by Emirate and service point
2	Insight into end to end capacity utilisation rates • Utilisation rates for SP intake, back office capacity, etc.
3	Proactive identification of potential bottlenecks and associated root causes: • Fees, photo, scanned documents, training.
4	Definition of analytically sound short and long-term solutions (e.g. setup of task forces to tackle short term backlogs, implementation of automated solutions).
5	Formulation of action plans for cross functional implementation of solutions.
6	Monitoring of implementation by departments and/or external vendor project teams.
7	Implementation of solutions for complex problems and management of crisis: • Application verification, card processing, production backlogs.
8	Documentation and transfer of knowledge to Emirates ID and external vendors teams: • Back office and typing centre manuals, staff training.

Having stabilised the enrolment processes in conjunction with departments and vendor project teams, the SSO undertook efforts to assess and streamline the card delivery processes. The drivers for the SSO engaging in this study were sizeable card delivery backlogs that developed due to increases in average daily enrolment volume and inefficient delivery mechanisms (see Figure 1.11).

The project commenced with conducting an analysis of existing card production and delivery backlogs and identifying bottlenecks and redundancies. Next, a benchmarking study was conducted to compare Emirates ID's card delivery processes versus comparable ID card programs in other countries (see Figure 1.12). The results of the benchmarking study and process assessment were used to define improvement objectives, as well as delivery and operational improvement

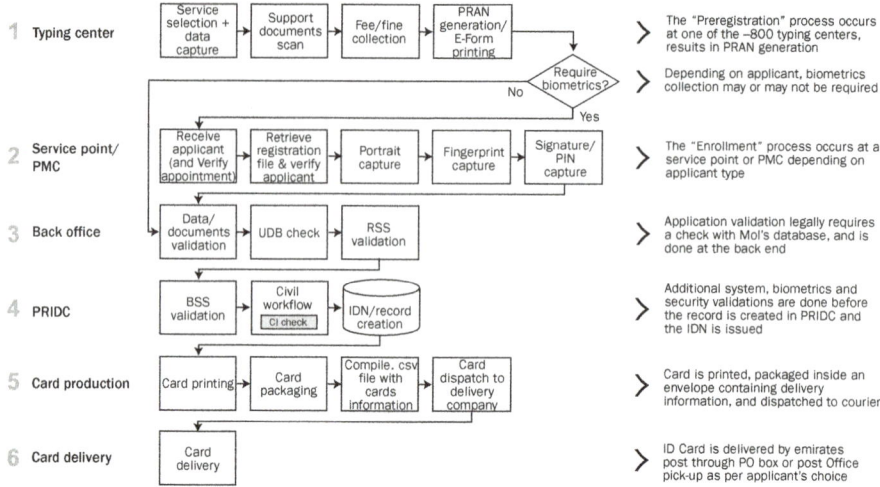

Figure 1.10 High-level overview enrolment process steps

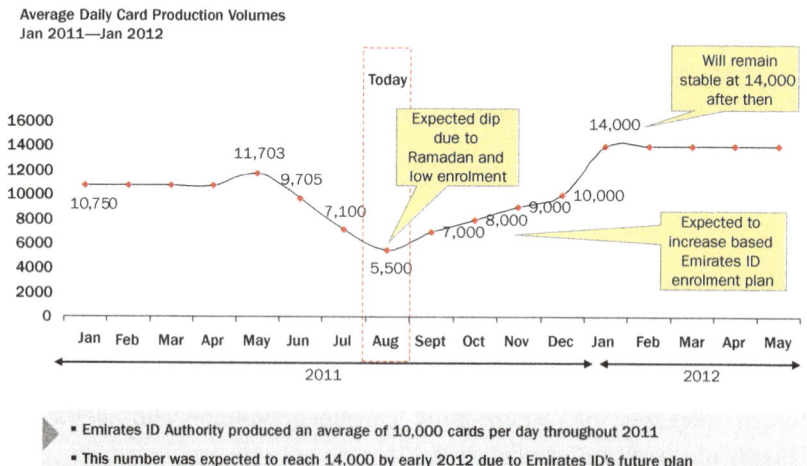

Today, Emirates ID prints on average 7,000 cards per working day and increased this number to 14,000 by 2012

- Emirates ID Authority produced an average of 10,000 cards per day throughout 2011
- This number was expected to reach 14,000 by early 2012 due to Emirates ID's future plan for 2012 that would ensure enrolment and consequently, card production

Figure 1.11 Card production volumes (current state)

requirements. The last step in this project was assessing the benefits and implications of each potential card delivery solution and selecting vendors that could support the roll-out of these potential solutions.

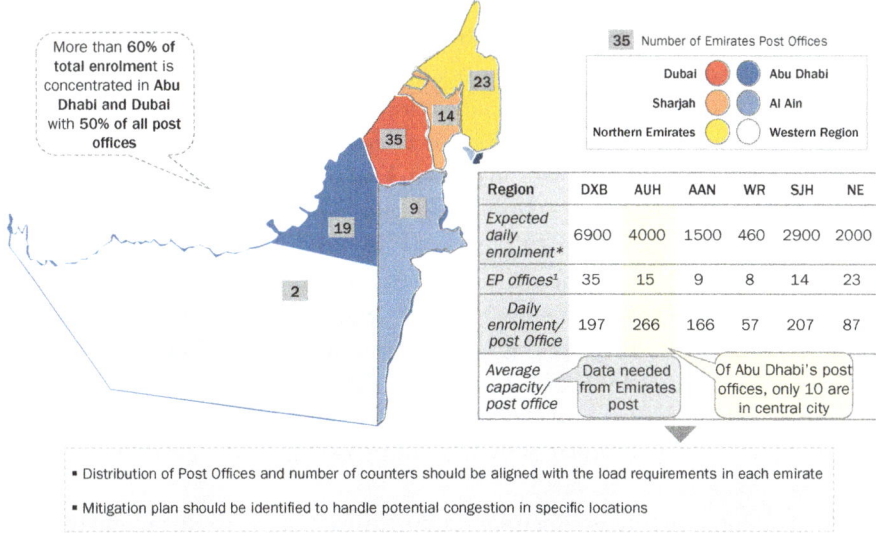

Figure 1.12 Emirates post pick-up locations

The outcome of the project was a complete revamping of Emirates ID's delivery approach from a 'delivery' to a 'pick-up' model. This enabled Emirates ID to greatly streamline its delivery processes and costs, thereby eliminating existing backlog levels, as well as reducing the risks of future backlogs. In addition, the selected approach also enabled the authority to leverage the scale and existing infrastructure of Emirates Post* for its card delivery needs. However, despite the process revamp, Emirates ID continued to face issues with card pick-up, as many customers were slow to retrieve their ID cards or choose to not pick up their cards at all.

One of the final projects undertaken as part of the strategy was the improvement of the performance of the call centre, which had been outsourced to two external vendors, each having ownership for a distinct subset of services. Given that the call centre served as the primary post-enrolment touch point for customers, its efficient operation was essential to upholding service delivery commitment to customers and, in turn, maintaining high customer satisfaction levels.

---

*Emirates Post is a government organisation in the United Arab Emirates established to provide postal and shipping services across the country and around the world.

As a first step, the authority conducted a thorough review of existing operations across both vendors by analysing the scope of their services, IT infrastructure, and inbound-outbound call volumes. In parallel, customer service teams worked to define the critical call centre functions required to better serve customers (e.g., general enquiries, application tracking, issue resolution and so on). These analyses helped provide an accurate view of the feasibility of continuing relationships with existing partners versus seeking alternate options to meet call centre service needs (see Figure 1.13).

Based on these initial analyses, the authority defined a set of options for the future call centre environment (e.g., single vs. multiple points of contact, call routing options, and so on), and obtained agreement from leadership teams for the desired option. Next, a road map was developed to transition from a multipoint to a single point of contact environment, identify key operational activities, and work with the customer service team to shortlist vendors.

Having evaluated numerous vendors, the team selected a company for a pilot project aimed at testing the adequacy of its IT infrastructure and customer service team. The last step in ensuring a seamless transition from a multipoint to single-point of contact solution was conducting a detailed mapping of the future IVR solution (level one and two services) and outlining all system and human intervention requirements.

Today, Emirates ID uses two call centres:Etisalat and Emaratech, both handling different loads and offering different services

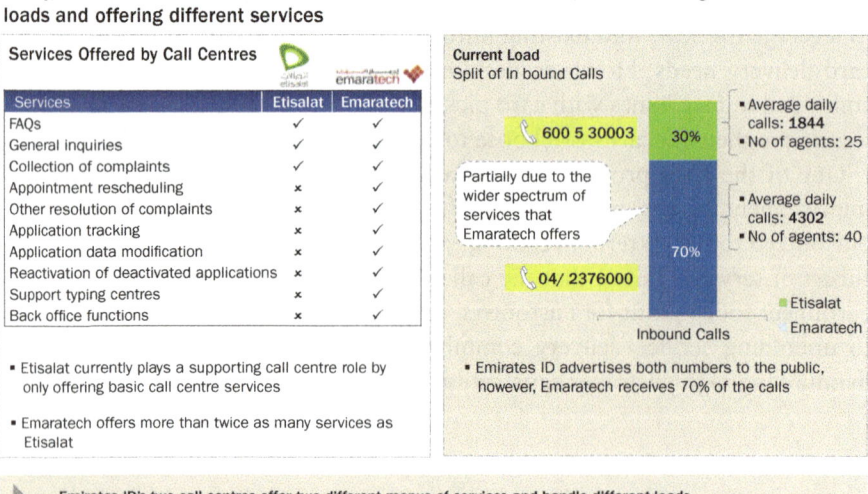

Services	Etisalat	Emaratech
FAQs	✓	✓
General inquiries	✓	✓
Collection of complaints	✓	✓
Appointment rescheduling	✗	✓
Other resolution of complaints	✗	✓
Application tracking	✗	✓
Application data modification	✗	✓
Reactivation of deactivated applications	✗	✓
Support typing centres	✗	✓
Back office functions	✗	✓

Current Load
Split of In bound Calls

600 5 30003 — 30%

Partially due to the wider spectrum of services that Emaratech offers

04/ 2376000 — 70%

- Average daily calls: 1844
- No of agents: 25

- Average daily calls: 4302
- No of agents: 40

Inbound Calls — ■ Etisalat Emaratech

- Etisalat currently plays a supporting call centre role by only offering basic call centre services

- Emaratech offers more than twice as many services as Etisalat

- Emirates ID advertises both numbers to the public, however, Emaratech receives 70% of the calls

Emirates ID's two call centres offer two different menus of services and handle different loads of calls causing some varying levels of service

Figure 1.13 Current call centre vendor services and volumes

The IVR for Emirates ID's unified call centre number will be updated to reflect the below structure and will results in the automatic routing of calls to either Emaratech or DNATA call centre agents

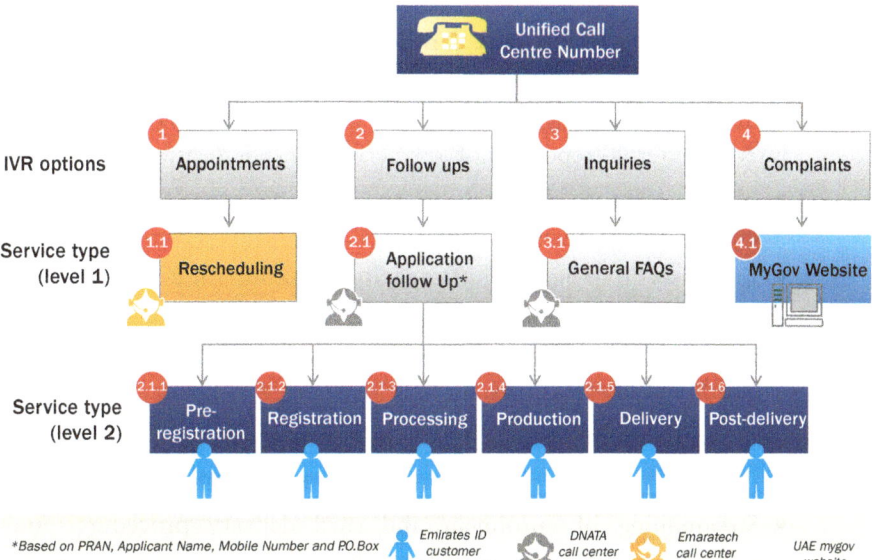

Figure 1.14 Proposed new IVR call routing tree

Although the identified solution was still in its early phase during the writing of this article, initial results were positive and showed that full implementation of a single point of contact was a feasible, practical solution that, nonetheless, required the implementation of a customised CRM solution that aligned with the authority's customer service needs and data requirements.

4. Strategy impact on Emirates ID

The new strategy had a tangible positive impact on the authority's overall performance. Not only was this visible in improved performance indicator results but also in terms of greater leadership team collaboration and employee engagement in adopting the new strategy. Below are some key areas that are viewed to have been positively affected:

1. Enabled enrolment acceleration acted as an enabler for the acceleration of enrolment rates via the facilitation of the implementation of re-engineering enrolment processes and active monitoring of enrolment processes. There was an:

- Increase in the Population Register from less than two million to more than five million.
- Increase in average daily enrolment from under 5,000 per day to more than 12,000 per day.

2. Performance tracking and reporting allows regular monitoring and internal-external reporting of key performance indicators:

- Improvement in KPI results and a more prevalent culture of accountability for KPI results.
- Emphasis on monitoring of capacity utilisation rates at registration centres.
- Elimination of root causes/drivers for bottlenecks across the enrolment process steps.

3. Increased customer centricity:

- Ongoing efforts to monitor and improve the customer experience.
- Streamlining of enrolment and card delivery procedures and improvement of existing and introduction of new customer touch points.

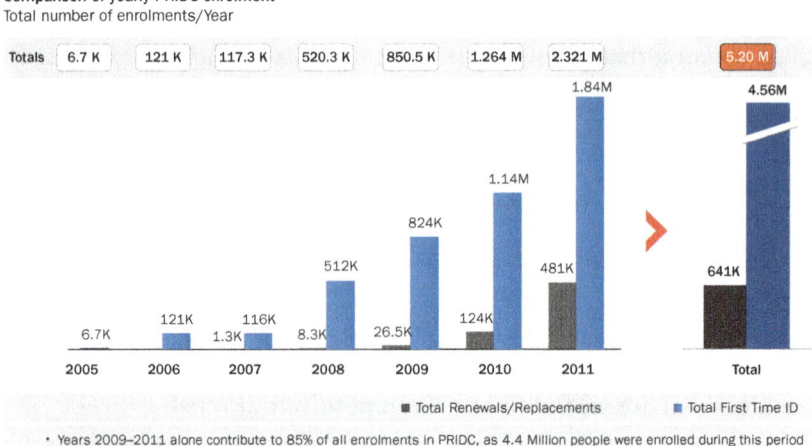

Comparison of yearly PRIDC enrolment
Total number of enrolments/Year

- Years 2009–2011 alone contribute to 85% of all enrolments in PRIDC, as 4.4 Million people were enrolled during this period

Figure 1.15 Annual enrolment volumes

Utilisation of PMC sites fluctuates across sites, with some centres operating at 106% capacity while others are underutilised, operating at 12% capacity

Emirate	EIMASS Site	Daily capacity	Average Daily Intake	Intake as % of Capacity	Status
Abu Dhabi	Abu Dhabi	1,260	849	67%	🟢
	Mussaffah	1,806	1,127	62%	🟢
	Madinat Zayed	218	189	90%	🟡
	Ghayathi	168	269	160%	🔴
Dubai	Knowledge village	210	110	52%	🟢
Sharjah	Al Baladiya	551	109	20%	⬜
	Al Ghabiba	673	81	12%	⬜
	Al Sinaiya	393	151	38%	⬜
	Airport freezone	84	41	49%	🟢
	Al Thaid	126	49	96%	🟡
	Khorfakan	71	64	89%	🟡
	Diba Al Hosn	84	23	27%	⬜
UAQ	UAQ	84	54	64%	🟢
Furjairah	Furjairah	259	30	12%	⬜
	Diba Al Furjairah	126	84	67%	🟢

🟡 Operation close to capacity ⬜ Underutilised site 🟢 Capacity and intake aligned 🔴 Capacity exceeded

Figure 1.16 **Utilisation rates monitoring dashboard**

4. Skills development led to acceleration in the development of technical and soft skills of the strategy department team and department project managers:

- Taking an unstructured problem and selecting appropriate information to analyse the issues while keeping in mind the final objectives and not getting lost in the details.

- Independently developing and applying a well-thought-out comprehensive plan.

- Conducting research that has predefined objectives, gathered by leveraging appropriate secondary sources and supported by meaningful data.

- Adeptness at quantitative and qualitative analytical techniques.

- Rapidly executing project tasks by focusing on core issues and effectively managing deadlines with limited guidance.

- Summarising/synthesising relevant findings and implications into well-structured presentations.

Call centre pilot: accomplishments and challenges

15 fully trained call centre agents operational six days a week from 8 am – 8 pm

Challenges

693 customer cases created with all required information, in the last three days

4.8 minutes on average required to handle each customer case

98% of inbound calls to the call centre were answered within 30 seconds

20% of applicants calling the call centre received an SMS with their case number

Accomplishments

400 calls received per day due to technical limitations from Etisalat

Figure 1.17 Call centre accomplishments and challenges

5. Lessons learned

5.1 Functional and technical aspects contributing to success

A key functional aspect that led to the success of the new strategy development was the setup of the SSO's working model as a task force supporting stakeholders from across departments. Such a model enabled key initiatives to benefit from a combination of the analytical and problem-solving skills of the SSO team and the technical know-how of vendor and department project teams. Furthermore, such a model required the close involvement of representatives from each department, thereby leading to greater buy-in, commitment, and accountability across the organisation. In addition, setting up the Strategy Support Office as a task force provided each department's leadership with specialised resources from which they could seek assistance whenever they faced challenges in setting up an initiative or needed assistance in resolving issues faced by an initiative. The end result was that stakeholders were empowered to manage initiatives internally but also had access to specialised resources to help them successfully meet commitments that may be at risk.

Some of the positive technical aspects of the project were systematic upfront planning, structured project management processes, and inclusion of data to support decision-making. The benefits of these technical elements were visible in the impact on and positive outcomes of initiatives where time was invested to adequately pre-plan activities, map project phases, and define mechanisms to measure progress (e.g., daily reporting, automated online reports). Furthermore, these initiatives successfully leveraged data as a means of quantifying the current situation, constantly assessing the impact and outcomes of implementing improvements (e.g., backlog management).

5.2 Functional and technical challenges faced

While there were many positives from the strategy development project, there were also some functional and technical challenges faced by the project team: One of the key challenges was the extremely limited pool of internal resources available for integration into initiative project teams. This limitation led to the small number of available skilled resources being stretched across too many initiatives, consequentially leading to a decrease in the use and output of resources.

5.3 Management success stories

A key management success was the involvement of and oversight by the senior management team in the Strategy Support Office. This involvement reassured employees and vendors that the leadership team was engaged and interested in the day-to-day operations of the organisation and was willing to hold people accountable for delays or underperformance. A key operational element that contributed to this management success was the implementation of the bi-weekly management meeting as a forum for reporting successes and quick wins, as well as escalation of issues that needed support from management and risks that could potentially delay execution of the strategy.

5.4 Management challenges

One of the challenges management faced was too much of the senior management teams' time consumed in identification and implementation of solutions for tactical and/or operational issues, thereby shifting the management teams' focus from strategic governance to operational governance, and in turn leading to an overburdened steering committee maintaining oversight across almost all initiatives (i.e., biweekly management team meeting). To prevent such a situation in the future, the mid-management team was asked to take on a more proactive, independent role in issue resolution, whereas the senior management team ensured that these individuals were empowered to make decisions and given the room to learn from their mistakes.

Although a key accomplishment for Emirates ID has been the rapid pace at which it has rolled out key initiatives to close gaps identified during the strategy development phase, this approach has also had an indirect downside: the adoption of such rapid change has led to a lack of prioritisation, thereby leading to competing priorities for many parts of the organisation and incomplete implementation of some initiatives. Underlying drivers for this have been the limited adherence by management to predefined project plans, calling for a phased roll-out of initiatives, and the lack of a central program management office responsible for maintaining an integrated master plan and enforcing disciplined issue management processes and standardised project reporting.

5.5 Best practices adopted

As mentioned earlier, Emirates ID was one of the first UAE federal entities to successfully implement the customised strategy development framework developed by the UAE Prime Minister's office. Incorporation of best practices into Emirates ID's strategy development efforts not only ensured compliance with the PMO's guidelines, but also enabled it to adopt a leading framework that incorporated both qualitative and quantitative analysis, ensuring a linear linkage of high-level mandate, strategy, and operational activities.

Conclusion

The 2010–2013 strategy of Emirates ID yielded successful outcomes in the first two years of its execution. Key factors that contributed to the overall strategy success were leadership commitment, a vigorous and dynamic management mindset, clear vision of expected outcomes, simplified thinking models, communication, and changes to management plans.

Government programs normally tend to be clear to some extent of their vision and high-level outcomes when they are first announced. However, our experience indicates that many formulated government strategies do not capture or take into account the overriding vision due to the complexity or nature of the program, hence facing significant difficulties to articulate requirements in clear terms of objectives, associated initiatives, and the setting up of precise and measurable KPIs. Therefore, we envisage the failure rate in the government sector to be high.

National identity card programs have been implemented around the world, and due to their complex nature, are no exception. Throughout this article, we have attempted to outline the role of the strategy in supporting the success of such programs. Although the limitation of a single-case study could be an obvious shortcoming, we aimed to document our accumulated knowledge and disseminate it to the field of research and practice, anticipating that it will allow a better understanding of government practices, share viewpoints and contribute to the advancement of government systems. Overall, the content of this article provides deep insights into Emirates ID's strategy development. It presents some useful information to practitioners, specifically those in the same field.

In addition to what we reported earlier, there are also important aspects that need management consideration: One is the need for a periodical review of the strategy. It is important that reviews be conducted regularly to help shape and adjust department activities. We propose that the framework should incorporate an assessment of key success factors (KSFs) into the strategy review process. The perceived benefit is that it would help create a stronger link between strategic objectives and operational activities, given that key success factors are the most important strategic enablers for an organisation to realise its strategic objectives (de Wit and Meyer, 2010). Furthermore, an assessment of KSFs would also help in the identification of concrete actions that could be fed directly into the departmental operating plans, thereby limiting the loss of any momentum in the future.

The review process should measure performance via three key quantitative and qualitative indicators, as depicted in Figure 1.18. However, to create a stronger link between strategic objectives and operational activities, a thorough assessment should be conducted for each KSF included in the strategy.

Another aspect for consideration is that government practitioners may use Moore's (2002) maturity lifecycle (depicted in Figure 1.19) to support their strategy development and review cycles. Moore's work places emphasis on improving shareholder value as the key driver for management decisions and to achieve and sustain competitive advantage.

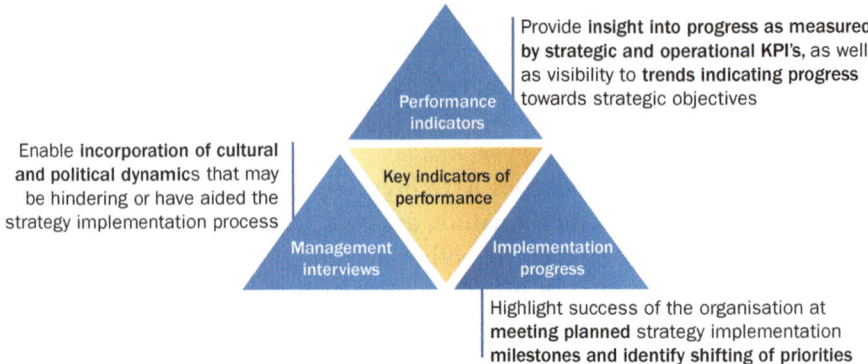

Provide **insight into progress as measured by strategic and operational KPI's**, as well as visibility to **trends indicating progress** towards strategic objectives

Enable **incorporation of cultural and political dynamics** that may be hindering or have aided the strategy implementation process

Highlight success of the organisation at **meeting planned** strategy implementation **milestones and identify shifting of priorities**

Figure 1.18   **Key indicators of performance**

The lifecycle consists of five stages: emergent, growth, maturity, decline, and end of life. The maturity lifecycle also overlays four value disciplines, described by Treacy and Wiersema (1995), namely:

1. Operational excellence: Differentiation based upon productivity and ultimately cost.

2. Customer intimacy: Differentiation based upon matching customer expectation with offer fulfilment.

3. Product leadership: Superior design and performance.

4. Category renewal of disruptive innovation: Create a new source of competitive advantage.

The model is recognised as a tool for analysing the dynamic evolution of organisations, thereby affecting the strategy development exercise. The model should be used to broaden the outlook on opportunities for growth. The key factor here is 'focus'. Management needs to decide on a single 'value discipline' and then construct its organisation around it (Treacy and Wiersema, 1995). Choosing one discipline to master does not mean discarding the others, but it means that the organisation needs to focus its energy and assets on a single discipline to achieve sustainable success: Each discipline demands a distinct strategy, organisational model with its own structure, processes, information systems, management systems, and culture.

Figure 1.19 Moore's maturity lifecycle

In principle, government strategies need to focus on: (a) core processes and competencies; (b) amplifying development through creativity and innovation; and (c) managing citizen relationships and promoting engagement. These are the keys to securing value creation, thereby creating more citizen-centred strategies. In short, and to effectively set successful strategies, organisations need to analyse these lifecycles and map them against their existing positions and growth plans. Developed strategies need to create value as quickly, effectively, efficiently, and accurately as possible. This is the only way we can speak the language of the globalised world we live in today.

References

1. Ahoy, C. (1986) 'Strategic Planning'. Iowa State University. Available at: *http://www.fpm.iastate.edu/worldclass/strategic_planning.asp*
2. Al-Khouri, A. M. (2011) 'Re-thinking enrolment in Identity Card Schemes'. *International Journal of Engineering Science and Technology* 3 (2): 912–925.

3. Al-Khouri, A. M. (2012) 'Projects Management in Reality: Lessons from Government Projects'. *Business and Management Review* 2 (4): 1–14.

4. Bryson, J. M. (2011) 'Strategic Planning for Public and Non-profit Organisations: A Guide to Strengthening and Sustaining Organisational Achievement'. San Francisco: Jossey-Bass.

5. DeWit, B. and Meyer, R. (2010) 'Strategy, Process, Content, Context'. London: Cengage Learning EMEA.

6. Lipman, F. D. and Lipman, L. K. (2006) 'Corporate Governance Best Practices: Strategies for Public, Private, and Not-for-profit Organisations'. Hoboken, NJ: John Wiley and Sons.

7. Moore, G. (2002) 'Living on the Fault Line'. New York: Harper Collins.

8. Moore, G. (2008) 'Dealing with Darwin: How Great Companies Innovate at Every Phase of their Evolution'. New York: The Penguin Group.

9. Moore, G. (1999) 'Crossing the Chasm'. New York: Harper Collins.

10. Steiner, G. A. (1997) 'Strategic Planning'. New York: The Free Press.

11. Treacy, M. and Wiersema, F. (1995) 'The discipline of market leaders: Choose your customers, narrow your focus, dominate your market'. Reading, MA: Perseus.

Customer relationship management: proposed framework from a government perspective

Abstract: Customer Relationship Management (CRM) has grabbed the attention of both practice and research in the past decade, developing into an area of major significance. The focus of the CRM concept is to build a long-term and value-added relationship for both an organisation and customers. Governments – although considered late followers compared to the private industry – have been showing growing interest in CRM systems recently to help public and government agencies track and manage relationships with their constituents. In this article, we review the existing literature to provide an understanding of the field. We also present a proposed CRM framework based on practice. The proposed framework is envisaged to act as a practical management tool that provides a holistic overview of implementation phases, components of each phase, and associated critical success factors.

Keywords: *Customer Relationship Management, CRM framework, CRM limitations and challenges*

1. Introduction

In terms of today's governments good governance is determined by citizen satisfaction. Satisfaction is a term frequently used in private sectors, referring to the measurement of how a product and/or a service supplied by a firm meets or surpasses customers expectations (Soudagar et al., 2011). It is generally defined as 'the number of customers, or percentage of total customers, whose reported experience with a firm, its products, or its services ratings exceeds specified satisfaction goals' (Farris et al., 2010).

Most importantly, governments across the world are finding themselves obliged more than ever to get closer to citizens and create systems that meet expectations. Citizens are demanding the same convenient service in the public sector that they are, for the most part, used to enjoying in the private sector. Providing satisfactory services determines the way any given government is viewed by the citizens and the rest of the world. According to today's international perspectives, a progressive government includes its citizens on the path of progress keeping in touch with their needs and requirements and, more importantly, providing a willing ear to hear their voices. The focus of attention is on enhancing the role of civil society and on the growing demand for good governance (Tembo, 2012).

The relationship with and management of citizens and residents – the customers and participants of the government – thus rank very high in government work. Governments need to focus on delivering high-quality, customer-centric, and integrated government services with the key strategic enabler being citizen-centric service (Al-Khouri, 2012). They need to place citizens at the heart of governmental work, and promote change in the government business to operate in a more citizen-centric way.

Throughout the public sector, initiatives to 'reinvent government'* have elevated customer service and satisfaction to new priorities (Osborne and Gaebler, 1992). Within Arab countries, for example, and with the advent of the Arab Spring, there is a shift in governments' mindsets to rethink and reform social services with social inclusion and 'user involvement' as driving forces in quality improvement. Customer relationship management (CRM) is becoming a top priority in government business to help agencies achieve their goals of developing models of service that are more responsive, more citizen-centric, and more efficient.

However, practical research states that there is a considerable state of confusion in the academic and managerial literature about what is meant

---

*The term 'reinventing government' emerged as an outcome of Clinton-Gore administration's interagency task force to reform and streamline the way the United States federal government functions. The concept, however, has been in practice in the private sector since the mid-1980s, where it is more commonly referred to as 'business process re-engineering', or simply 're-engineering'. Today, these terms are, for the most part, used interchangeably, although some in government still prefer to use the term 'reinvent' as opposed to 're-engineer'.

by CRM; despite heavy investment by organisations in CRM, there is extensive reporting of CRM's failure to achieve anticipated results (Frow and Payne, 2009). The existing literature also points out that despite the increasing interest in CRM in government work, the adoption of CRM systems by government agencies – and particularly by e-government programs – is slow (Pan et al., 2006). A lack of strategic focus is recognised as one of the prime reasons for CRM failures. We also note that little has been written about how the public sector might use CRM principles to improve service delivery.

Considering the strategic context of CRM systems in organisations, this article emphasises the need for a conceptual framework to guide management in the implementation of CRM systems. We present a framework that is based on findings from existing research and practice. The proposed framework is envisaged to further support its practical applications in implementing successful CRM systems. It may act as an effective guiding tool for management to provide a holistic overview of the implementation phases, the components of each phase, and the associated critical success factors.

This article is structured as follows: First we provide a short review of the literature in section 2, with a focus on defining the CRM concept. We also attempt to provide a summary of the most common critical success factors for CRM implementation as reported in the literature. Then in section 3, we examine a generic CRM framework to underpin its components. In sections 4 and 5, we explore the role of technology in CRM programs, its potential benefits, and identified limitations. In section 6, we explore the role CRM technology as a contemporary communication and service delivery enablers. In section 7, we briefly draw on cloud-based CRM solutions as an alternative implementation approach. In section 8, we review some complexities associated with CRM programs. Then, in section 9, we outline our research and development methodology. We finally present the proposed framework and explain its components in section 10.

2. Customer Relationship Management (CRM)

Dowling (2002) suggests that CRM had its origins in two unrelated places: One was in the United States, where it was driven by technology in connection with customer-based technology solutions; the other was in Scandinavia and Northern Europe to support business-to-business marketing in connection with the Industrial Marketing and Purchasing (IMP) Group that has been instrumental in developing knowledge about the nature and effects of building long-term, trust-based relationships with customers.

The concept of CRM evolved because it places an emphasis on understanding customer needs and then solving problems or delivering benefits that create demonstrable customer value (Dowling, 2002). The role of information technology is important in this style of CRM, as it is designed to support – rather than drive – the customer relationship. The types of relationship that develop here are often deep and meaningful, both for the firm and the people involved (Dowling, 2002).

All in all, CRM has developed into an area of undeniable significance in less than two decades (Frow and Payne, 2009). According to a recent IDC report, the CRM industry revenues exceeded US$19 billion in 2011 (IDC, 2012). The huge scale and scope of the inter- and intra-organisational changes involved in CRM led Kotorov (2003) to assert that CRM was the third most significant revolution in the organisation of business after the invention of the factory in 1718 and the introduction of the assembly line into factory production in 1913.

The literature identifies two streams of research to form the theoretical foundation of a CRM concept (Agrebi, 2006): a strategic stream (relationship marketing) and a technological stream related to the information systems (Triki and Zouaoui, 2011). Relationship marketing (RM) is more of a strategy designed to foster customer loyalty, interaction, and long-term engagement (Mintzberg, 1994). It is normally designed to develop strong connections with customers by providing them with information directly suited to their needs and interests and by promoting open communication. This approach often results in increased word-of-mouth activity, repeat business, and a willingness on the customer's part to provide information to the organisation. Almost all organisations practise aspects of RM (Frow and Payne, 2009). However, while CRM emphasises the integration of processes across different functions, customer management is concerned with the tactical aspects of CRM implementation that relate to the management of customer interactions, including the use of tools such as campaign management, sales force automation, Web-enabled personalisation, and call centre management (see also Figure 2.1).

In fact, Crosby and Johnson (2001) identify customer relationship management as a business strategy that multiplies the use of technology

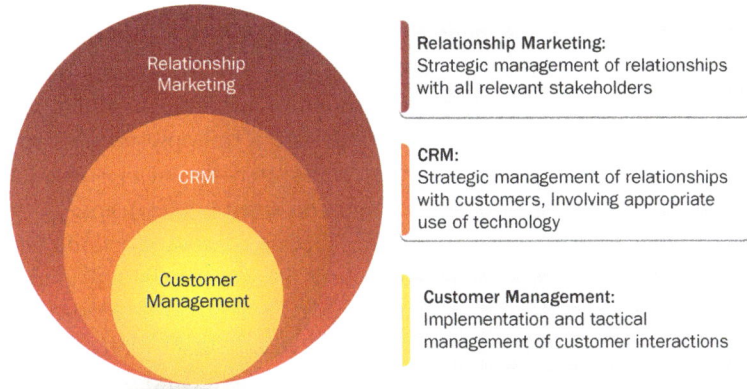

Relationship Marketing:
Strategic management of relationships
with all relevant stakeholders

CRM:
Strategic management of relationships
with customers, Involving appropriate
use of technology

Customer Management:
Implementation and tactical
management of customer interactions

Figure 2.1   Relationship marketing, CRM and customer management

Source: Frow and Payne (2009)

and includes it in all processes to create retention and loyalty over time. In general terms, the focus of the CRM concept is to build a long-term and value-added relationship for both business and customers. From this perspective, let us review some definitions to clarify the term:

- 'Coherent and complete set of processes and technologies for managing relationships with current and potential customers and associates of the company, using the marketing, sales and service departments, regardless of the channel of communication' (Chen and Popovich, 2003);

- 'Customer Relationship Management (CRM) is a business strategy to select and manage customers to optimise long-term value. CRM requires a customer-centric business philosophy and culture to support effective marketing, sales, and service processes. CRM applications can enable effective Customer Relationship Management, provided that an enterprise has the right leadership, strategy, and culture' (Thompson, 2002);

- 'To improve service and retain customers, CRM synthesises all of a company's customer touch points' (Yu, 2001);

- 'Good customer relationship management means presenting a single image of the company across all the many channels a customer may use to interact with the firm, and keep a single image of the customer that is shared across the enterprise' (Berry and Linoff, 2000).

These and other definitions suggest three key dimensions associated with the term CRM (see also Figure 2.2). First, CRM is about business strategy focused around the customer. It is built around customers to manage beneficial relationships through acquiring information on different aspects of customers. Nevertheless, a full commitment of the organisation's staff and management is essential for an effective CRM implementation to best serve customers and satisfy their needs.

Second, CRM is about the business processes that support and enable the interaction between a business and its customers. All business processes that involve both direct and indirect interaction with customers should be analysed and assessed (Mendoza et al., 2007). Third, there are the technology and management dimensions. The technology part refers to computing capabilities that allow organisations to improve their understanding of customer behaviour, develop predictive models, build effective communications with customers, and respond to those customers with real-time, accurate information (Chen and Popovich, 2003). The other part of the dimension involves continuous corporate change in culture and processes, as CRM requires comprehensive change in the organisation and its people.

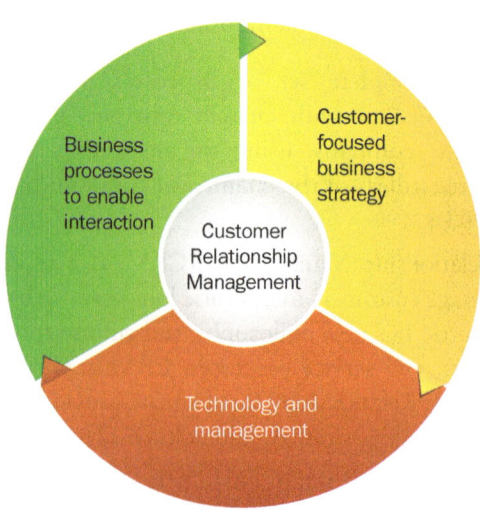

Figure 2.2   The components of CRM

The role of the strategy is considered a critical component (Newell, 2003; Fayerman, 2002; Starkey and Woodcock, 2002; Rigby et al., 2002; Crosby, 2002; Winer, 2001; Yu, 2001). The CRM strategy must define the 'what' and 'how' elements of the organisation's intents that should aim to create a 'single integrated view of customers' and a 'customer-centric approach' to address their customers' needs (Roberts-Witt, 2000). The strategy process should also attempt to learn more about customers' needs and behaviour to develop stronger relationships with them and create public value (Coltman et al., 2003). Building the right type of relationship is the key to performance improvement (Coltman, 2006).

The literature emphasises that a deeper understanding of the dynamics of the CRM triangle components is important (Johnson, 2004; Teece et al., 1997). Successful CRM case studies indicate the collective role of the technical, human, and business capabilities (Coltman, 2006). The reason for this is that each capability is within an intricate organisational system of interrelated and interdependent resources (Coltman, 2006). This resource-based view (RBV) has been subject to criticism (see, for example, Day et al., 2002; Harris, 2001; Jaworski and Kohli, 1993). This argument is based on the fact that such a narrow perspective lacks sound operational criteria to distinguish important capabilities in addressing a wider view of associated elements. Nonetheless, the type of emerging thinking from the existing literature points to the need to thoroughly focus on human, technical, and business capabilities as main determinants of successful CRM systems. Each element needs to be interpreted in the context of business.

One of the problems with the CRM performance literature to date is that there is a temptation to be normative about the pursuit of market orientation based on the identification of certain CRM capabilities (Coltman, 2006). The field of practice exploited this shortcoming to ruthlessly market CRM products as a technical solution. This narrow view has contributed to the significant percentage of failure within their CRM projects implementations.

Reports from Gartner Group and Meta Group had three very striking findings: more than 50 per cent of CRM implementations are viewed as failures by the customer, 55-75 per cent of CRM implementations fail to meet their objectives, and customers usually underestimate the costs of CRM implementations by 40-75 per cent (Coltman, 2006). A 2009

Forrester research study found that 47 per cent of CRM implementations fail – many due to a lack of CRM strategy and user adoption (Simon, 2010). The study elaborated on some of the problems experienced during CRM implementation. The problems most commonly cited by executives are depicted in Figure 2.3.

Existing research points out a number of success factors that organisations need to keep in mind when aligning their strategy with objectives and managing people performance to ensure execution and results. Tables 2.1 and 2.2 provide a summary of previous studies on CRM success factors. Most of these are later incorporated into our proposed framework in section 10.

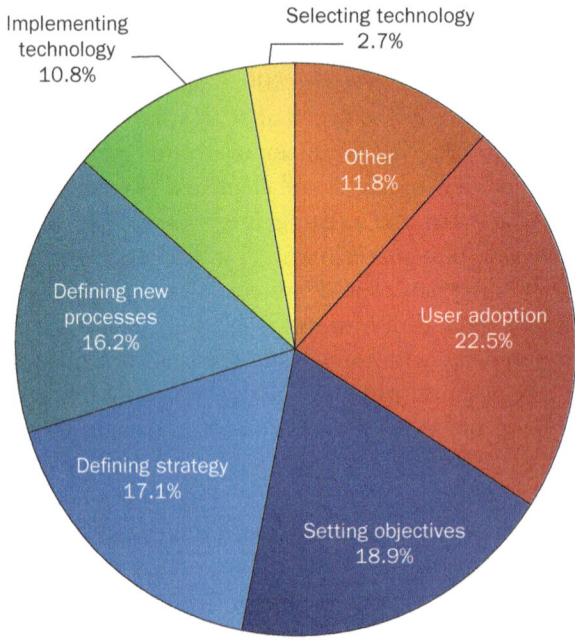

Figure 2.3 CRM platform implementation problems

Source: Simon (2010).

Table 2.1 CRM success factors

King and Burgees (2007)	Chalmeta (2005)	Da Silva and Rahimi (2007)	Pan and et al. (2007)	Alt and Puschmann (2007)	Saloman et al. (2005)	Mendoza et al. (2006)
Top management support	Awareness among management	CRM philosophy	Evolution path	Evolution path	Top management commitment	Senior management commitment
Communication of CRM strategy	Defining vision and objectives	Project mission	Timeframe	Timeframe	Change in corporate culture	Creation of multi disciplinary team
KM capabilities	Creation of committee	Top management commitment	Organisational redesign	Organisational redesign	Significant customer data	Objective definition
Willingness to share data	Official appointment of coordinates	Project schedule and plan	Reorganisation	System architecture	Clearly defined CRM processes	Interdepartmental Integration
Willingness to change process	Development and approval of the project plan	Clint consultation	Minimize customisation	Change management	Sufficient resources	Communication the CRM strategy to the staff
Technological readiness	Monitoring to control time slippage	Connectivity	Time and budget management	Top management support	Understanding of customer behaviour	Staff commitment
Cultural change / customer orientation	Prevent resistance to change	Skilled personnel	Customer involvement		Extensive IT support	Customer information management

Table 2.1 CRM success factors (Cont'd)

King and Burgees (2007)	Chalmeta (2005)	Da Silva and Rahimi (2007)	Pan and et al. (2007)	Alt and Puschmann (2007)	Saloman et al. (2005)	Mendoza et al. (2006)
Process change capabilities	Motivate staff	Technical tasks	No culture conflict			Customer service
System integration capabilities	Measure the degree of participation/ Asses the results	Client acceptance	Use of the CRM system managers			Sales automation
		Monitoring and Feedback	Management involvement			Marketing automation
		Communication				Support for operational management
		Troubleshooting				Customer contact management
		BPS and software configuration				Information systems integration

Source: Adapted from Almotairi (2009)

Table 2.2 CRM success factors

Mankoff (2001)	Eid (2007)	Wilson et al. (2002)	Goodhue et al. (2002)	Croteau and Li (2003)	Siebel (2004)	Chen and Chen (2004)	Roh et al. (2005)
Establish measurable business goals	Top management support	Gain champ	Top management support	Top management support	Clear communication of strategy	Champion process fit leadership and internal marketing	Process fit
Align business and IT operations	Organisational culture	Ensure market orientation	Vision	Technological readiness	Back-office integration	Business-IT alignment	Customer information quality
Get executive support up front	Developing a clear CRM strategy	Define approval procedures which allow for uncertainty	Willingness to change process	KM capabilities	Software customisation	System integration	System support
Let business goals drive functionality	Clear project vision/scope	Gain board awareness of strategic potential of IT	Willingness to share data			KM	Efficiency
minimise customisation by leveraging out-of-the-box functionality	Benchmarking	Identify need for business system convergence				Culture/structure change	Customer satisfaction
Use trained, experienced consultants	Employees acceptance	Organise around customer					Profitability

Table 2.2 CRM success factors *(Cont'd)*

Mankoff (2001)	Eid (2007)	Wilson et al. (2002)	Goodhue et al. (2002)	Croteau and Li (2003)	Siebel (2004)	Chen and Chen (2004)	Roh et al. (2005)
Actively involve end users in solution design	CRM software selection	Address culture change					
Invest in training to empower end users	Integration with other systems	Involve users in system design					
Use a phased rollout schedule	Training	Manage IT infrastructure					
Measure, monitor, and track	Realistic CRM implementation schedule	Leverage models of best practice					
	Enterprise performance metrics for CRM	Rapid strategy/action loop to experiment					
	Personalisation	Prototype new processes					
	Customer orientation	Manage for delivery of benefits					
	Data mining	Design for flexibility					

Source: Adapted from Almotairi (2009)

3. CRM components

A critical point to remember here is that while governments are not driven by profit or revenue generation, they are driven by the demand to create public value (Al-Raisi and Al-Khouri, 2008). The creation of public value – although not the same as the generation of revenue – is analogous. The components of CRM remain the same for the private as well as the government sectors with a difference in outlook and drivers. Figure 2.4 depicts a citizen service framework that should be part of any CRM system.

The citizen-centric government seeks to provide higher levels of customer satisfaction. In such models, it treats itself as a service delivery organisation with integrated citizen (customer) transactions and benefit delivery, providing multiple interfaces of interaction. The multiple interfaces of interaction translate into multiple channels of transaction. The government is no longer confined to four walls and filing cabinets. CRM capabilities are viewed as a key enabler for such transformations. They are examined in the framework of their components. The common key components of CRM framework are illustrated in Figure 2.4. These are:

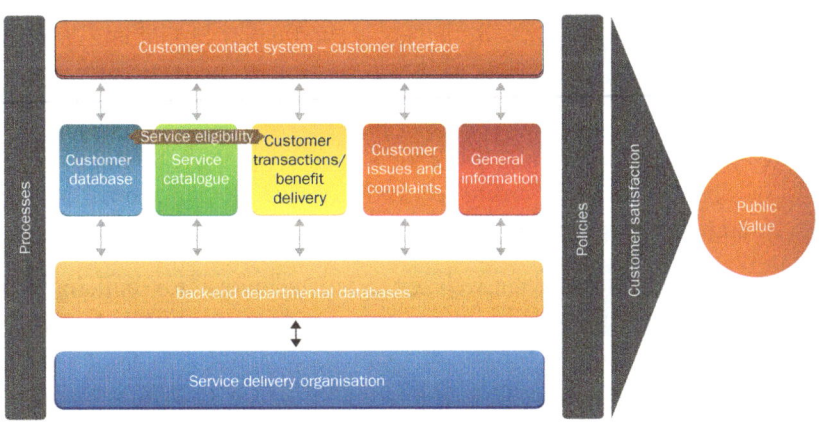

Figure 2.4 Citizen service framework

1. A customer interface/contact system. This enables the customers to interact with the government department delivering a set of services.

2. Customer database. This is a crucial component that ensures personalisation of services.

3. Service catalogue. This is a detail of the services provided by the organisation. For government departments, the service list is complemented by benefit and eligibility lists, detailing who can be accorded what service.

4. Customer transactions. These provide valuable information about customers' interactions with the government department.

5. General information. This is a repository of all information about the government department delivering the services which can aid the customer. Data includes information on new services, social message campaigns, and promotion of the government department.

6. Processes. A set of workflows, procedures, and related metrics for delivering services. Measurement of these provides valuable input for the analysis of the efficiency of the service delivery.

7. Policies. These guidelines provide the rules and scope of engagement with the customers.

8. Back-end databases. These give crucial information from internal sources of the government department providing a response to customer requirements.

9. Service delivery organisation. This is the human capital engaged in the delivery of the services to the customers: at the front interacting with the customer, at the office back-end, and field service delivery.

10. Customer satisfaction. This is the output from the CRM measured by different surveys, onsite service delivery KPIs and results on the final outcome of public value creation for the government.

CRM is definitely an extensively executed technique for building an organisation's relationships along with citizens, government agencies, and third-party organisations. It requires making use of modern technology to arrange, speed up and also connect business processes and those elements for promoting customer support along with technical support.

4. The allure of CRM technology

In practice, CRM requires efficient, integrated business systems, as it imposes an organisation-wide discipline to develop a single image of the customer shared across the enterprise. The main categories of benefits from CRM-based work systems touch on multiple performance dimensions such as operational, managerial, strategic, infrastructure, and organisational (Davenport et al., 2002; Shang and Seddon, 2002). Examples of these benefits include: improved management decisions, improved customer service, and increased development opportunities (Freeman and Seddon, 2005). Other benefits include: increased productivity from headcount reductions and other process efficiencies; integration of processes, data and technology; consistency and standardisation of processes and information; business measurement and reporting; personalised and responsive service to customers; and increased sales activities (Freeman and Seddon, 2005). We elaborate on the main ones here:

- Enhanced customer experience. CRM, backed by the correct technology, serves to improve the customer experience with service delivery organisation. Each of the technologies previously mentioned contribute to this end. Voice-over IP improves communications by unifying the networks over which communication is delivered. This not only reduces the costs, but also enables contact agents to be located at various geographic locations. Voice communication backed by instant messaging and chat provide increased customer participation in the service delivery.

- Customer convenience. By providing multiple channels of communication, a customer gets a wide choice to interact with the Government Service Delivery Organisation and provides the ability to interact at his or her convenience.

- Better service access. Speech applications improve the IVR, enabling customers to interact with service databases directly and securely without having to wait for a personal contact.

- Transparency and comprehensive information. Web services and business integration platforms enable disparate databases to work together, providing comprehensive information drawn from different data sources. This allows for transparency in service delivery transactions where different oranisations are engaged in the delivery process and work on various applications for their business processes.

- Presence technologies and personalisation. A high level of personalisation in service delivery can be achieved by understanding the profiles of customers. RFID, smart cards, and point-of-sale systems provide information on the presence of a customer at a particular point. Knowledge about the presence of customer at different locations helps greatly in delivering the service requested by accessing the profiling of the customer.

- Queue management and cycle time reduction. These help ensure that unnecessarily long waiting times are avoided and quicker services are provided to customers.

- Connectivity. The importance of good networking and telecommunication systems cannot be understated in the delivery of services to the customer. This is true for an interaction of the customer with a government service delivery organisation via any channel of communication. Networking technology contributes to the reliability of service delivery.

- Analytics (improved management and decision-making). All through the process of service delivery, and throughout the service lifecycle, a huge amount of data is generated that leads to valuable information. Analysis of this information provides deep insights into the entire service delivery process itself and a better understanding of the customer. A good analytical system helps decision-making and improves business' organisation.

Although the potential value of CRM technologies is enormous, organisations have had difficulties in taking advantage of all the possibilities. One reason is that CRM systems require significant changes to existing practices and a significant amount of process development is required. For example, organisations need to think through what constitutes a CRM system and ensure alignment with the culture and environment they operate in. Also, supporting defined processes within a CRM system is often not as straightforward as it might appear and may require a reasonable amount of customisation. In other words, CRM technology will not provide all the benefits on its own. The right strategies and supporting processes need to be developed, and existing systems need to be tuned or re-engineered to support them. While this is not – as the saying goes – rocket science, it can be sufficiently involved to discourage the more casual user of CRM technology (Boarman, 2011).

5. The limitations of CRM technology

As with most available technologies, there is no single one that fits the requirements of a perfect customer relationship management system. Different components of customer relationships have different technological challenges in deployment. Integration (or non-integration) is a big challenge in effective CRM systems. Information not available in real time can lead to dissatisfaction. Tracking of applications for service requests is a prerequisite. The complexity is enhanced when we consider the multiple channels of communications available to interact with customers. A person might call on the phone to register a request, follow up with an e-mail, submit documents in person, receive an SMS to provide a status, visit the website for tracking the request, and so on. Such interactions and other complex interactions need to be well managed. There is no single tool that facilitates the complete set of these interactions. Integrated systems need to be deployed and managed to facilitate these communication channels.

Furthermore, back-end systems need to be updated to handle the service requests and reports need to be provided for monitoring the service level agreements (SLAs). The measurement of customer satisfaction and public value creation is another major challenge. Setting up improper metrics would lead to wrong analyses. A 2005 Bain Consulting study revealed that 81 per cent of senior leaders in 362 surveyed firms believed their organisation delivered 'superior customer service', yet only eight per cent of their customers agreed (Allen et al., 2005). The study refers to the problem as a 'customer service gap'. Others call it an example of falling into the trap of overestimating one's achievements and capabilities in relation to how others view them. The study also suggests that most leaders are out of touch with their organisation's customer experience and the engagement of their frontline employees.

Management needs to focus on realigning its goals and measures, systems and organisational structures to design the right customer experiences and deliver them flawlessly. They need to see their systems from the eyes of their customers, not their own. Measurements that lead to the correct portrayal of the goal's achievement should be selected. There is no technology that provides these metrics, though once set up, the systems can intelligently monitor them.

No CRM system can come up with bulletproof processes that will guarantee success. Processes are to be defined internally and managed by a qualified group of process analysts. Unless processes are in place, monitoring the effectiveness of the service delivery is not possible. Yet, despite the difficulties and associated challenges, several trends currently drive CRM programs at all levels of governments – the most important of which is related to streamlining government service delivery for an improved response. Government agencies can transform the way they deliver services to their citizens, achieve greater operational efficiency, and proactively improve the communities they serve.

6. CRM technology in the context of contemporary communication and service delivery channels

Existing practices of CRM associate it with electronic services (e-services) and electronic government (e-government) (Pan et al., 2006; Richter et al., 2004). e-Service refers to any service provided by any electronic means (e.g. Internet/website, mobile devices or kiosk). According to Grönlund (2005), e-service is a core component of e-government because it bridges the gap between the government administrators and citizens. Figure 2.5 shows e-service as one of the main participants in the e-government domain; the arrows indicate 'influence', the circles indicate 'domains of control', and the intersection of circles indicates 'transaction zones'. In a democratic government system, the triangular relations are vital where service delivery is one of the main interactions between public servants (administration) and citizens and businesses (civil society).

As such, there is a government attempt to create service-oriented architectures (SOA)* and develop a single window platform through which public services are provided on a 24/7 basis to allow citizens to electronically interact and transact with government agencies (Al-Khouri, 2012). Electronic services reflect three main components: the service provider, the channels of service delivery (i.e., technology), and the service receiver. For instance, public agencies are the service providers, while citizens and businesses are the service receivers.

---

*In software engineering, a Service-Oriented Architecture (SOA) is a set of principles and methodologies for designing and developing software in the form of interoperable services. SOA design principles are used during the phases of system development and integration to define how to integrate disparate applications for a Web-based environment and uses multiple implementation platforms. SOA is not just an architecture of services seen from a technological perspective, but the policies, practices, and frameworks by which we ensure the right services are provided and consumed.

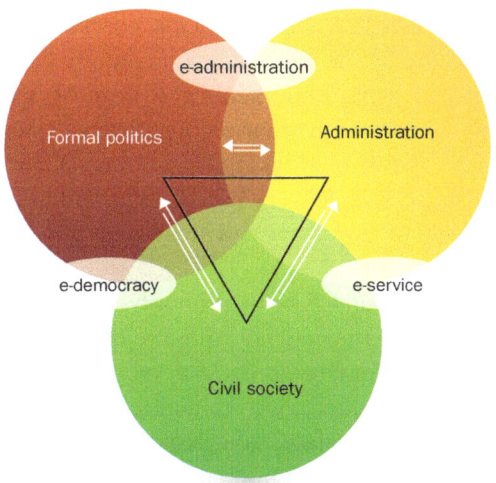

Figure 2.5 e-Service as a component of e-government

Source: Grönlund (2005)

The channel of service delivery is the third dimension of e-service and is about the enabling technologies – that is, the Internet as the main channel of e-service delivery while other classic and traditional channels are also considered such as telephones, call centres, public kiosks, mobile phones, televisions, over-the-counter service, and postal mail service. The extent of Internet-enabled CRM includes electronic CRM (E-CRM), mobile CRM (M-CRM), and ubiquitous CRM (U-CRM) (Chang and Wu, 2009).

- Electronic CRM (E-CRM): A concept derived from e-commerce. It uses intranet, extranet, and Internet environments (Reponen, 2003). There are major differences between CRM and E-CRM. From a customer contact perspective, CRM is contact made through the retail store, phone, and fax. E-CRM uses all traditional methods in addition to Internet, e-mail, wireless, and PDA technologies. E-CRM is geared more towards the front-end, which interacts with the back-end through the use of ERP systems, data warehouses, and data marts;

- Mobile CRM (M-CRM) uses wireless networks as the medium of delivery to the customers (Camponovo et al., 2005);

- Ubiquitous CRM (U-CRM), also referred to as Virtual CRM (V-CRM) uses virtual reality to create synergies between virtual and physical channels and reach a very wide consumer base. However, given the newness of the technology, most companies are still struggling to identify effective ways into virtual reality (Goel and Mousavidin, 2007).

Channels through which companies can communicate with their customers are growing by the day. Such contemporary interactions have the new dimension of 'virtual interactions' instead of 'traditional front-desk interactions'. The strength of virtual interaction is dominated by the e-service's existence and its quality. In fact, citizens or businesses usually choose a channel of service delivery based on its suitability and level of expertise. The blend of any service will be determined in relation to demand.

However, thinking about CRM in primarily technological terms is a confusion that many governments fail to figure out. However, we cannot deny the huge role for technology in CRM development. As an example of technology in CRM, communication technology could be seen to play a major role in customer interface. A customer could interact with the government service delivery organisation using multiple channels of contact and through various types of public e-services – from simple information dissemination to highly sophisticated automated e-service (see Figure 2.6).

Complementing communication technology, information technology provides process automation, data integrity, and security with much-needed confidentiality for the customer. This enhances the trust and convenience levels for the customer. With the widespread usage of mobile phones and technological advances, citizen service channels have

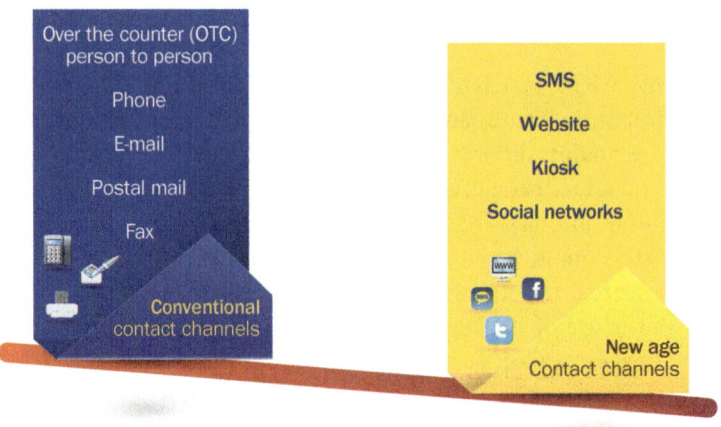

Figure 2.6   Transitioning in contact channels

become numerous in their deployment. Internet portals as a single window of communication and interaction has been implemented by many governments (Bukhsh and Weigand, 2011; Lenihan, 2008; Monga, 2008). Different departments are engaged with citizens through one standardised interface of Web portals (see Figure 2.7).

Interactive voice response systems evolved with telephone systems and enabled information-based services with automation. Widespread computing has enabled the deployment of kiosk machines that have enabled government communication and service delivery channels to be located at places frequented by the citizens: shopping centres, recreational areas, and so on. Personal communications evolved in the impersonal domain of the Internet to secure identifiable personal interactions. This has resulted in instant messaging as a medium of communication; with e-mail, online chats, and so on.

Help desks and call centres have evolved to full-fledged contact centres, enabling citizens to communicate and interact using different mediums of communication. Governments have recently adopted social media networks such as Facebook and Twitter to bring communication with their citizens to a much more personal level vis-à-vis the impersonal abstract advertisements and brochures of the past.

Development in all these channels – combined with developments in mobility, network security and digital identity management – have transformed conventional government authorities into fully e-enabled governments. Figure 2.8 outlines these different channels in the context of citizen services in e-government scenarios.

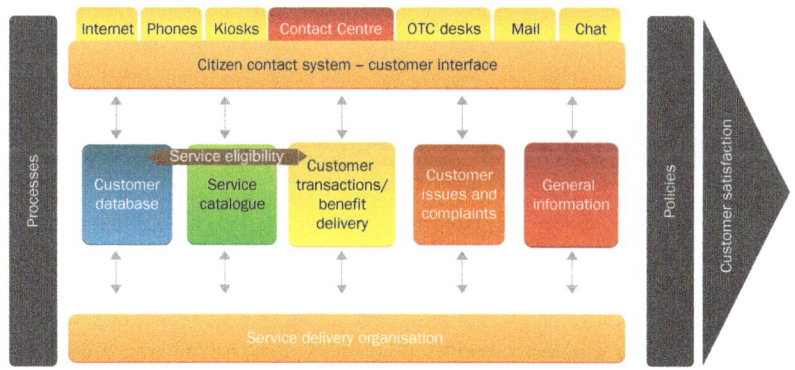

Figure 2.7 Customer citizen framework and CRM communication channels

CRM	Technology		
	Voice communications	Data communications	Networking and connectivity
Customer contact	– Voice over IP – IVR	– SMS – e-mail – Web portal – Kiosks	Unified networks and communications
Service catalog customer transactions issues and complaints general information	Speech applications	– CRM application – Web services – Service oriented architecture, and business process integration – Relational databases	Social networking
Backend data	Network connectivity and applications for business process automation		
Service delivery	Queue management – Presence technologies (RFID, PoS	CRM application for transaction management	Networks and logistics
Analytic	CRM Application tool with embedded analytic and business intelligence (data warehousing and data mining)		

Figure 2.8 CRM and communication technologies

7. Cloud-based CRM solutions

Today, more and more organisations are drifting away from physical locations of data and moving to Internet-based cloud-computing solutions. Cloud-based* technology simply means that the technology does not live in an IT-based environment and can be delivered through the Internet, which means that agents, supervisors, and executives can all access the same information in real time. SaaS and cloud CRM solutions have spurred the evolution of computing – with no more software

---

*Cloud computing is the use of computing resources (hardware and software) delivered as a service over a network (typically the Internet). There are three types of cloud computing: infrastructure, platform and software as a service (SaaS). Using SaaS, users can rent application software and databases. Cloud providers manage the infrastructure and platforms on which the applications run. Cloud computing relies on sharing of resources to achieve coherence and economies of scale similar to a utility (like the electricity grid) over a network. At the foundation of cloud computing is the broader concept of converged infrastructure and shared services.

installations, no infrastructure management, and no more upgrades to test. With SaaS and cloud CRM solutions, development and implementation can now be accomplished in a fraction of the time required for on-premise solutions (Bennett, 2010) (see also Figure 2.9).

According to Gartner Research, CRM went up from eight per cent in 2005 to 20 per cent of the market in 2008 (Bennett, 2010). Cloud-based CRM systems can be cost efficient as they can be managed, maintained and upgraded on a pay-per-use basis. Some enterprise CRM in cloud systems are real-time Web-based and need no additional interface installed. People may communicate on mobile devices to get these services. Customer experience and feedback are other ways of CRM collaboration and integration of information in corporate organisations to improve business' services (Chandrasekaran and Kapoor, 2010).

Ultimately, the cloud is a growing trend amongst businesses hoping to take advantage of the ability to host technology without having to maintain a cumbersome database, and the CRM industry is steadily coming into play in the cloud-based arena.

Generally speaking, governments around the world have been showing little interest in cloud-based solutions. There is much more preference to use, design, and develop CRM systems in-house. Nonetheless, the field of practice in the government sector shows the adoption of unfocused and various tools and techniques to manage and provide online services

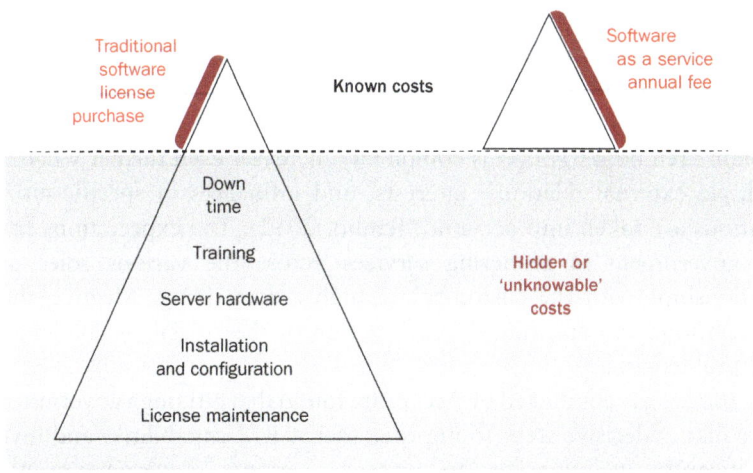

Figure 2.9 SaaS vs. on-premise total cost of ownership

Source: Özcanli (2012)

to citizens (Yarmoff, 2001). Unfortunately, the fragmented nature of their approach prevents such systems from being truly effective (Yarmoff, 2001). However, we believe that CRM technologies – in whatever form – will be used more than others by governments in the coming ten years. This is because CRM is not a technology but rather is a business strategy. However, there are different layers of complexity associated with CRM systems in government contexts that need to be looked at and carefully managed.

8. The complexity of customer relationships in government

A citizen-centric approach needs to be viewed as a basis for government work. Managing the ever-changing roles of the citizens with the government is paramount to conscientiously managing their relationship. In a private sector customer relationship, the management is relatively simpler. The customer of a private firm, to a reasonable extent, is clearly defined, and so are the services and products. The private sector, furthermore, is focused on an objective of higher revenue generation driven by higher customer satisfaction.

In comparison, the customer relationship in which a government is engaged is far more complex in nature. The citizen is as much part of the government as he or she is a customer. The citizen entity itself is dynamic, as one may play multiple roles: as a citizen, an employee, a service provider, and so on. These relations, in themselves, are a complex web of formal and informal interactions that are difficult to disentangle and explain (Tembo, 2012). This complexity increases even further when the multiple external relations, interests, and influences in specific citizen relations are taken into account (Tembo, 2012). The expectations from the government in rendering services across the various roles and relationships with the customer are highly demanding. Meeting these expectations and ensuring the satisfaction of the citizens is challenging and problematic.

A 2002 study conducted by Accenture found that although governments were taking decisive steps to improve their CRM capabilities and invest significantly in initiatives to improve services, governments were struggling to realise the benefits expected from developing modern CRM capabilities (Crook et al, 2002). Research indicates that many governments still have not been able to bridge the gap between the envisioned impact of CRM and their current experience (Gilbert et al., 2004; Kavanagh, 2007; Silva and Batista, 2007). Figure 2.10 illustrates the extent to which government agencies are considered to be lagging behind the private sector in developing intelligent customer interactions that are driven by customer intentions. Some have progressed to deliver multichannel interaction. Many have implemented little or no service automation.

Undoubtedly, this is a real challenge. How does a government department manage all its customers? How does it track all the transactions and understand the nature of relationship? How does it ensure that required services are delivered to the correct customer? How does the government track the satisfaction and, more importantly, measure this satisfaction and ensure a satisfied nation? The answers to all these questions lie in good customer relationship management. Good CRM is not just a technology but a set of best practices, good processes, and meaningful metrics to address citizens' needs. It is for this reason that we started writing this article and put forward a proposed framework to guide CRM implementation. The next section will explore the methodology used to construct our proposed framework, and we will present our framework in the subsequent section.

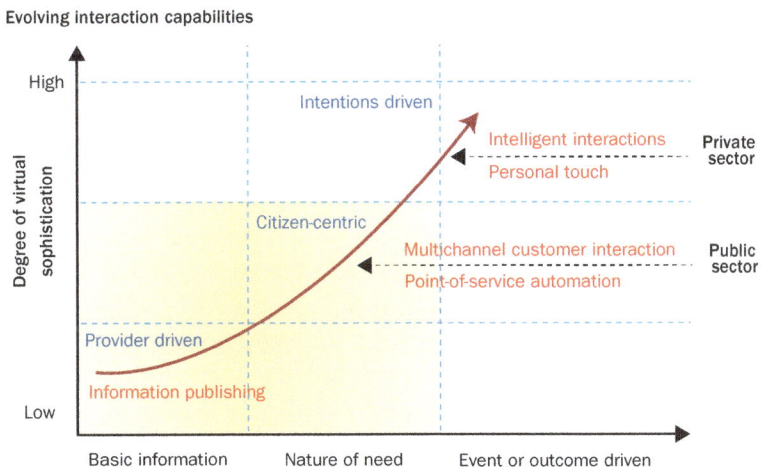

Figure 2.10 Interaction capabilities of the public and private sector

Source: Crook et al (2002)

9. Research and development methodology

The approach followed in this study was based on meta-analysis methodology. Meta-analysis refers to methods focused on contrasting and combining results from different studies in the hope of identifying patterns among study results, sources of disagreement among those results, or other interesting relationships that may come to light in the context of multiple studies (Glass 1976; Greenland and Rourke, 2008). Meta-analysis is argued to be the most important methodological innovation in the social and behavioural sciences in the last 25 years, developed to offer researchers an informative account of which methods are most useful in integrating research findings across studies (Hunter and Schmidt, 2004).

So far, the most often used meta-analysis has been in the literature review of quantitative (statistical) research that helps the general strength of the effect under different circumstances. More recently, meta-analysis has become more common in diverse research fields (DeCoster, 2009).

Meta-analysis can be a very useful method to summarise data across many studies, but it requires careful thought, planning, and implementation (Denyer and Tranfield, 2006). Not surprisingly, as with any research technique, meta-analysis has its advantages and disadvantages: one advantage is its objectivity, yet as with any research, its value ultimately depends on making some qualitative-type contextualisations and understanding of the objective data.

Our study focuses on the results of existing research and practice to construct our proposed conceptual CRM framework. It uses variables from different research studies and incorporates it in the framework design to enhance its practicality. We also use our experience in the field of government to further refine the concept and to reflect management needs. We aim to promote this framework to be used as a simplified management tool to support comprehension and, thereafter, successful CRM implementation in the government sector.

Our preference for this research methodology follows the recommendation of Denyer and Tranfield (2006), who argue that meta-ethnographers infer that it is possible to translate the findings of some studies into the terms of another to build higher-level constructs. Besides, if converged with qualitative research, approaches like meta-analysis can provide management with a means of creating actionable knowledge in the future (Davies et al., 2000; Noblit and Hare, 1988).

10. A proposed CRM framework

The implementation of CRM systems involves a great deal of complication and challenges. CRM is not a single technology that can be implemented with a magic trick to increase customer satisfaction. A CRM system is a comprehensive set of processes and tools backed by technology. Proper planning and a phased approach are recommended for the adoption of a CRM system. We propose a five-stage approach to deploy CRM systems (see Figure 2.11).

We incorporate these stages into a more comprehensive framework illustrated in Figure 2.11. The framework includes a number of additional factors and focuses on the representation of identified critical success factors in the literature to improve usability and application in practice. The framework is designed in three phases: pre-implementation, implementation, and post-implementation. Each phase has a number of critical success factors associated with it that need management attention. Figure 2.12 illustrates the framework.

In the first phase, organisations need to visualise and define the goals and objectives of a CRM system. This should set the desired expectations that should be translated later into an input to build CRM capabilities by balancing short-term impact with long-term strategy. This phase also includes creating a proposed design for the new system in the form of requirements, critical priorities, and overall system architecture. It should also list the CRM components and related technology components. The output of this phase will be used as input for the next phase.

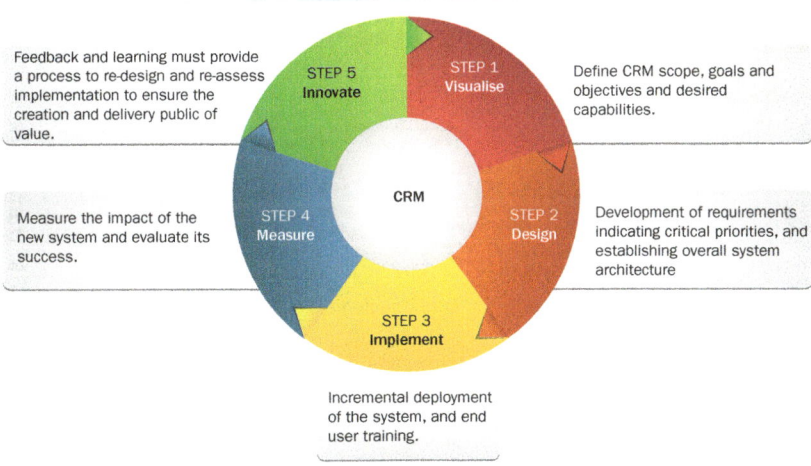

Figure 2.11 A five-stage approach to deploy a CRM system

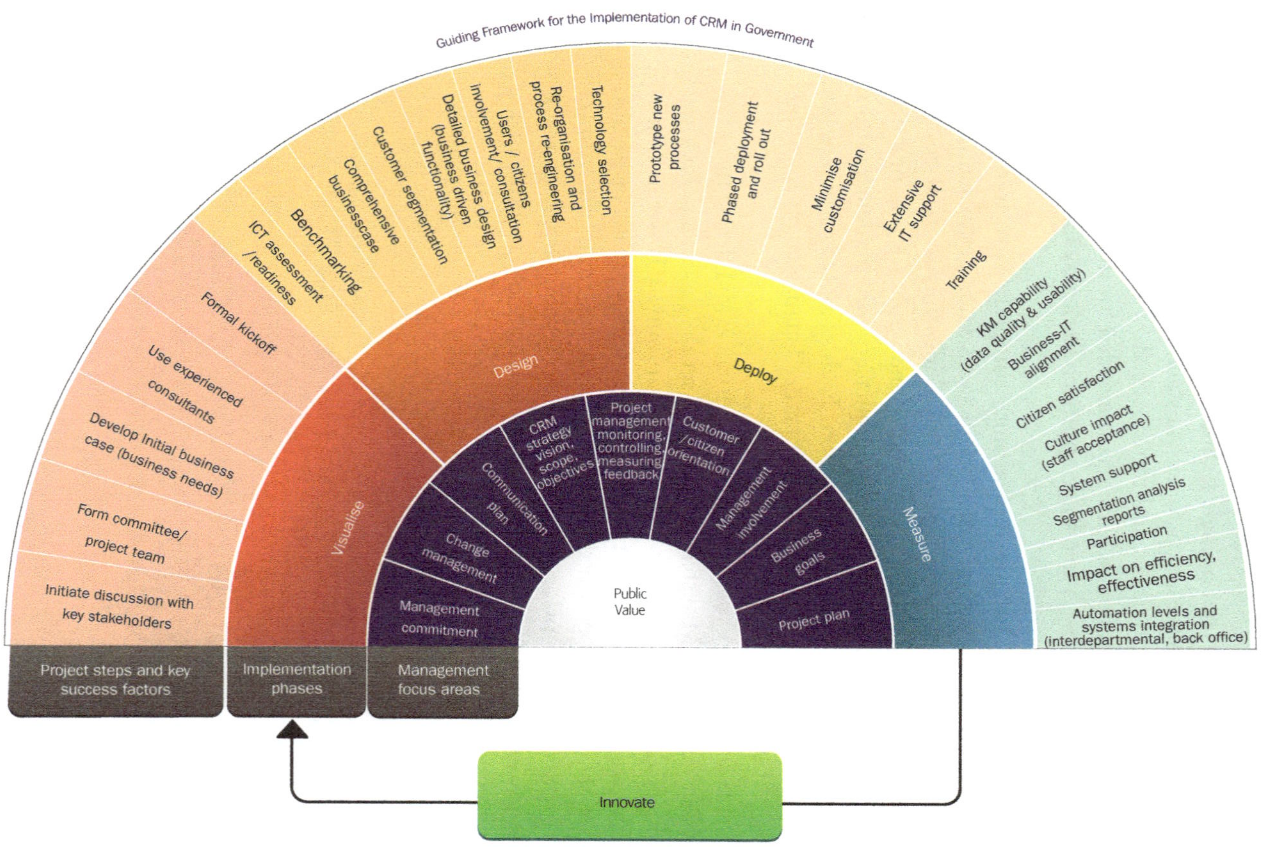

Guiding Framework for the Implementation of CRM in Government

Figure 2.12 CRM framework

The next phase is about the implementation of the system. The selected components are deployed, implemented, and the output and outcomes monitored. An incremental deployment of the CRM system provides a time-phased, measured roll-out.

The third phase, post-implementation, is basically about measuring the impact of the new system and evaluating its success in achieving set objectives and outcomes, and the stage is set for the next iteration. This iteration approach seeks to introduce innovation through a feedback and learning process to redesign and reassess its implementation. Innovation in technology would need to be considered to optimise the business process and the related customer experience. This phase ultimately represents the stage where governments attempt to create and deliver public value.

This framework design was also based on three dimensional streams: management focus areas, project steps and key success factors in relation to each of the implementation phases. Management focus areas include a set of necessary elements that need to be followed to promote the culture of result-based management. Result-based management is a management strategy that ensures that the processes, output, and services contribute to the achievement of clearly stated and measurable results that are aimed at improving the overall organisation performance. Such strategies encompass the following: empowering managers and holding them accountable; focusing on participation and partnership; developing support mechanisms; and creating a culture of information within the organisation. Result-based management has linked budget planning with strategic policy planning, thus moving from having an internal management focus to having an outward-looking citizen-centric orientation and managing for development of public value results.

The second dimension of project stages and key success factors include guiding processes and milestones that can be used in a CRM implementation program. These can be used to examine and determine the thoroughness of project plans. It is true that project planning is often ignored in favour of getting on with work. Most of the time, project plans are looked at from narrow and/or technical perspectives. The proposed elements are considered as fundamental building blocks for any sound project planning approach for a CRM program. Although the specifics of various programs may differ, many of the critical elements are envisaged to be similar.

The framework pays high attention to the critical success factors identified as part of our literature review. Thus, it has been developed around identified factors. Critical success factors attracted considerable attention in both practice and research as a means of supporting both planning and requirement analysis. When implementing CRM systems, they can be particularly effective in supporting planning processes, in communicating the role of information technologies to senior management, and in promoting structured analysis processes.

Customer segmentation is a recommended activity in the proposed framework. This is the practice of dividing a customer base into groups of individuals who are similar in specific needs and preferences such as age, gender, and other demographical and profiling attributes (Tsiptsis and Chorianopoulos, 2010). The concept of looking at groups of customers using the segmentation approach would allow government organisations to develop better understanding of responses to public needs (Barnett and Mahony, 2011). This should also allow them to better serve and maintain more effective relationships with their customers. The CRM solution must be designed to incorporate segmentation needs, with a focus on developing strategies that improve the delivery of vital citizen services – with a particular focus on how people move up the value tree (Wood, 2008).

However, a note to management: Critical success factors outlined in the framework should not be confused with success criteria. These are outcomes of a project or achievements of an organisation that are needed to consider the project a success or the organisation as successful (Boynlon and Zmud, 1984). Success criteria are defined with the objectives and may be quantified by key performance indicators (KPIs) (Friesen and Johnson, 1995). Critical success factors, on the other hand, are elements that are vital for a strategy to be successful.

A critical success factor should drive the strategy forward. It should make or break the success of the strategy. KPIs should be viewed as measures that quantify management objectives, along with a target or threshold, enabling the measurement of strategic performance. A successful critical success factor-led requirements analysis is likely to be supported by other, more concrete, techniques; or it is likely to involve management with a proactive, rather than reactive, organisational perspective (Boynlon and Zmud, 1984).

Conclusion

Customer relationship management is a requirement to reach out to customers proactively and provide personalised services. This will not only result in higher customer satisfaction but also, from a government perspective, dramatically improve its relationships with its customers through reorganising service delivery capabilities around customer intentions, thus creating real public value. When implemented well, CRM systems can meet the strategic objectives of the government in providing better services and a better quality of life for its citizens. A good CRM system should always be determined by its final outcome and should result in high public value creation.

A lack of understanding CRM systems has the potential to contribute to the failure of the whole initiative – especially when organisations view such systems from a purely narrow technological perspective or when they address CRM in a fragmented manner. Besides, there are different layers of complexity associated with CRM systems in government that need to be looked at and carefully managed. Governments need to concentrate on the needs of their citizens to achieve their goals of developing service architectures that are more responsive, more citizen-centric, and more efficient. To focus more on core competencies, governments may consider cloud-based CRM systems as a viable alternative compared to an in-house implementation model. However, concerns remain around the privacy, security, and sovereignty of data on such operating platforms (Chandrasekaran and Kapoor, 2010). Policy-makers need to strike the right regulatory balance to ensure flexibility, regulatory compliance and jurisdiction to allow cloud computing to perform in an efficient manner, driven by trust and confidence, and to inspire innovation (Chandrasekaran and Kapoor, 2010). More successful adoption cases of cloud-based CRM in the private sector is likely to encourage the public sector to follow suit.

In this article, we attempted to explore the field of CRM and relate the existing literature to the context of government. The framework provides management with a conceptual tool to guide management in the implementation of CRM systems. The contribution of our work resides in the design of the framework that integrates the five-stage CRM implementation, and the inclusion of relevant key success factors identified in the literature included in each phase. While the individual components of the framework can be viewed as being more abstract, the overall framework can be tailored to specific needs and settings.

Future work to test and refine the proposed framework is inevitable to evaluating and validating its practicality. For any framework to be successful, organisations need to understand and refine their own vision of how knowledge should be structured, communicated, and socialised within the organisation to influence results (Kellen, 2002). Then again, for governments to build successful CRM systems with new customer-oriented capabilities, they need to start constructing new ways of knowing their citizens.

References

1. Agrebi, M. (2006) 'Les Apports, Obstacles et Facteursclés de Succès d'une e-relation: Le Point de Vue des Fournisseurs de Solutions eCRM'. *Communications 5ème journée nantaise de recherche en e-marketing*, cited in: Triki, A. and Zouaoui, F. (2011) 'Customer Knowledge Management Competencies Role in the CRM Implementation Project'. *Journal of Organizational Knowledge Management*: 1–11.
2. Al-Khouri, A. M. (2012) 'e-Government Strategies: The Case of the United Arab Emirates'. *European Journal of ePractice* 17: 126–150.
3. Allen, J., Reichheld, F. F., Hamilton, B. and Markey, R. (2005) 'How to achieve true customer-led growth'. Bain and Company, Inc. Available at: *http://www.bain.com/Images/BB_Closing_delivery_gap.pdf*
4. Almotairi, M. (2009) 'A Framework for Successful CRM Implementation'. European and Mediterranean Conference on Information Systems July 13–14, Crowne Plaza Hotel, Izmir.
5. Al-Raisi, A. N. and Al-Khouri, A. M. (2010) 'Public Value and ROI in the Government Sector'. *Advances in Management* 3 (2): 33–38.
6. Alt, R. and Puschmann, T. (2004) 'Successful Practices in Customer Relationship Management'. *Proceedings of the 37th Hawaii International Conference on System* Science: 1–9.
7. Barnett, C. and Mahony, N. (2011) 'Segmenting Publics'. *National Co-ordinating Centre for Public Engagement. Economic & Social Research Council*. Available at: *http://www.publicengagement.ac.uk*
8. Bennett, T. (2010) 'Cloud CRM: Ready, Steady, Roll It Out'. Available at: *http://www.crmbuyer.com/story/69085.html?wlc=1289278204& wlc=1289626348*
9. Berry, M. and Linoff, G. (2000) 'Mastering Data Mining: The Art and Science of Customer Relationship Management'. New York: John Wiley and Sons.
10. Boarman, R. (2011) 'Benefits of CRM'. Mareeba Consulting. Available at: *http://www.mareeba.co.uk/blog/2011/02/crm-benefits-lead-management. html*
11. Boynlon, A. C. and Zmud, R. W. (1984) 'An Assessment of Critical Success Factors'. *Sloan Management Review* 25 (4): 17–27.
12. Camponovo, G., Pigneur, Y., Rangone, A. and Renga, F. (2005) 'Mobile Customer Relationship Management: An Explorative Investigation of the Italian Consumer Market'. *Proceedings of The Fourth International Conference on Mobile Business (ICMB 2005)*, Sydney, 11–13 July.
13. Chalmeta, R. (2006). 'Methodology for Customer Relationship Management'. *Journal of Systems and Software* 79: 1015–1024.
14. Chandrasekaran, A. and Kapoor, M. (2010) 'State of Cloud Computing in the Public Sector – A Strategic analysis of the business case and overview of initiatives across the Asia Pacific'. Frost and Sullivan. Available at: *http://www.frost.com/prod/servlet/cio/232651119*
15. Chang, W. L. and Wu, Y. X. (2009) 'A Framework for CRM E-Services: From Customer Value Perspective'. In Sharman, R., Raghu, T. S. and Rao, H. R. (Eds.) *Exploring the Grand Challenges for Next Generation E-Business,*

Lecture Notes in Business Information Processing: 235–242 Berlin Heidelberg: Springer-Verlag. Available at: *http://mail.tku.edu.tw/wlchang/LNBIP-520235.pdf*

16. Chen, I. J. and Popovich, K. (2003) 'Understanding customer relationship management (CRM): People, process and technology'. *Business Process Management Journal* 9 (5): 672–688.

17. Chen, I. J. and Popovich, K. (2003) 'Understanding customer relationship management (CRM) People, process and technology'. *Business Process Management Journal* 9 (5): 672–688.

18. Chen, Q. and Chen, H. M. (2004) 'Exploring the Success Factors of e-CRM Strategies in Practice'. *Journal of Database Marketing & Customer Strategy Management* 11 (4): 333–343.

19. Coltman, T. (2007) 'Why build a customer relationship management capability?', *Journal of Strategic Information Systems* 16 (3): 301–320.

20. Coltman, T., Devinney, T. M. and Midgley, D. (2003) 'Strategic Drivers and Organizational Impediments to eBusiness Performance: A Latent Class Assessment'. AGSM Working Paper Series.

21. Coltman, T. R. (2006) 'Where Are the Benefits in CRM Technology Investment'? *Proceedings of the 39th Hawaii International Conference on System Sciences* Available at: *http://ro.uow.edu.au/cgi/viewcontent.cgi?article=1237&context=infopapers*

22. Crook, P., Simmonds, A. and Rohleder, S. J. (2002) 'CRM in Government: Bridging the Gaps'. Accenture. Available at: *http://www.accenture.com/SiteCollectionDocuments/PDF/crm_bridging.pdf*

23. Crosby, L. (2002) 'Exploding some myths about customer relationship management'. *Managing Service Quality* 12 (5): 271–277.

24. Crosby, L. A. and Johnson, S. L. (2001) 'Technology: Friend or Foe to Customer Relationships'. *Marketing Management* 10 (4): 10–11.

25. Croteau, A. M. and Li, P. (2003) 'Critical Success Factors of CRM Technological Initiatives'. *Canadian Journal of Administrative Sciences* 20 (1): 21–34.

26. Da Silva, R. and Batista, L. (2007) 'Boosting government reputation through CRM'. *International Journal of Public Sector Management* 20 (7): 588–607.

27. DaSilva, R. V. and Rahimi, I. D. (2007) 'A Critical Success Factors model for CRM Implementation'. *International Journal of Electronic Relationship Management* 1 (1): 3–15.

28. Davies, H. T. O., Nutley, S. M. and Smith, P. C. (2000) 'What Works? Evidence-Based Policy and Practice in Public Services'. Bristol: Policy Press.

29. Day, G. S. and Van den Bulte, C. (2002) 'Superiority in Customer Relationship Management: Consequences for Competitive Advantage and Performance'. Cambridge: Marketing Science Institute.

30. Day, G. S. and Hubbard, K. J. (2002) 'Customer Relationships Go Digital'. *Business Strategy Review* 14 (1): 17–26.

31. Denyer, D. and Tranfield, D. (2006) 'Using qualitative research synthesis to build an actionable knowledge base'. *Management Decision* 44: 213–227.

32. Dowling, G. (2002) 'Customer Relationship Management in B2C Markets, Often Less is More'. *California Management Review* 44 (3): 87–104.

33. Eid, R. (2007) 'Towards a Successful CRM Implementation in Banks: An Integrated Model'. *The Service Industries Journal* 27: 1021–1039.
34. Farris, P. W., Bendle, N. T., Pfeifer, P. E., and Reibstein, D. J. (2010) 'Mrketing metrics: The definitive guide to measuring marketing performance' (2nd edition). Upper Saddle River, NJ: Pearson.
35. Fayerman, M. (2002) 'Customer Relationship Management'. *New Directions for Institutional Research* 113: 57–67.
36. Freeman, P. and Seddon, P. B. (2005) 'Benefits from CRM-Based Work Systems'. Available at: *http://is2.lse.ac.uk/asp/aspecis/20050017.pdf*
37. Friesen, M. and Johnson, J. A. (1995) 'The Success Paradigm: Creating Organizational Effectiveness Through Quality and Strategy'. New York: Quorum Books.
38. Frow, P. and Payne, A. (2009) 'Customer Relationship Management: A Strategic Perspective'. *Journal of Business Market Management* 3 (1): 7–27.
39. Gilbert, D., Balestrini, P. and Littleboy, D. (2004) 'Barriers and benefits in the adoption of e-government'. *International Journal of Public Sector Management* 17 (4): 286–301.
40. Glass, G. V. (1976) 'Primary, secondary, and meta-analysis of research'. *Educational Researcher* 5: 3–8.
41. Goel, L. and Mousavidin, E. (2007) 'CRM: Virtual Customer Relationship Management'. *Database for Advances in Information Systems* 38 (4): 56–58.
42. Goodhue, D. L., Wixom, B. H. and Watson, H. J. (2002) 'Realizing business benefits through CRM: Hitting the right target in the right way'. *MIS Quarterly Executive* 1 (2): 79–94.
43. Grabner-Kraeuter, S. and Gernot, M. (2002) 'Alternative Approaches Toward Measuring CRM Performance'. 6th Research Conference on Relationship Marketing and Customer Relationship Management, Atlanta, June 9–12.
44. Greenland, S. and O'Rourke, K. (2008) 'Meta-Analysis'. In: Rothman, K. J., Greenland, S. and Lash, T. L. (eds.) *Modern Epidemiology* 3rd edition: 652–682. Philadelphia, PA: Lippincott Williams and Wilkins.
45. Harris, L. C. (2001) 'Market Orientation and Performance: Objective and Subjective Empirical Evidence from UK Companies'. *Journal of Management Studies* 38 (1): 17–43.
46. Hunt, S. and Morgan, R. M. (1995) 'The comparative advantage theory of competition'. *Journal of Marketing* 57 (7): 1–15.
47. Hunter, J. E. and Schmidt, F. L. (2004) 'Methods of Meta-Analysis: Correcting Error and Bias in Research Findings'. Sage.
48. IDC (2012) 'The Fight for CRM Applications Market Leadership Gets Tighter'. International Data Corporation (IDC). Available at: *http://www.idccom/getdoc.jsp?containerId=prUS23539412*
49. Jaworski, B. J. and Kohli, A. K. (1993) 'Market orientation: antecedents and consequences'. *Journal of Marketing* 57 (7): 53–70.
50. Johnson, J. (2004) 'Making CRM Technology Work'. *British Journal of Administrative Management* 39: 22–23.
51. Kavanagh, S. C. (2007) 'Revolutionizing Constituent Relationships: The Promise of CRM Systems for the Public Sector'. Government Finance Officers Association. Available at: *http://www.gfoa.org/downloads/CRM.pdf*

52. Kellen, V. (2002) 'CRM Measurement Frameworks'. Available at: *http:// www.kellen.net/crm_mf.pdf*

53. King, S. F. and Burgees, T. F. (2007) 'Understanding success and failure in customer relationship management'. *Industrial Marketing Management* 34(4): 421–431.

54. Kotorov, R. (2003) 'Customer Relationship Management: Strategic Lessons and Future Directions'. *Business Process Management Journal* 9 (5): 566–571.

55. Mankoff, S. (2006) 'Ten Critical Success Factors for CRM: Lessons Learned from Successful Implementations'. White Paper. Available at: *http://www. oracle.com/us/products/applications/siebel/051291.pdf*

56. Mendoza, L. E., Marius, A., Perez, M. and Griman, A. C. (2007) 'Critical success factors for a customer strategy'. *Information Software Technology* 49: 913–945.

57. Mintzberg, H. (1994) 'The Rise and Fall of Strategic Planning'. New York: Prentice Hall.

58. Newell, F. (2003) 'Why CRM Doesn't Work'? Princeton, New Jersey: Bloomberg Press.

59. Noblit, G. W. and Hare, R. D. (1988) 'Meta-Ethnography: Synthesizing Qualitative Studies'. London: Sage.

60. Osborne, D. and Gaebler, T. (1992) 'Reinventing Government: How the Entrepreneurial spirit is transforming the public sector'. Addison-Wesley Publishing.

61. Özcanli, C. (2012) 'A proposed Framework for CRM On-Demand System Evaluation'. Master of Science Thesis Stockholm, Sweden. KTH Royal Institute of Technology Publication Database. Available at: *http://kth.diva-portal.org*

62. Pan, S. L., Tan, C. W. and Lim, E. T. K. (2006) 'Customer Relationship Management (CRM) in e-Government: A Relational Perspective'. *Decision Support Systems* 42 (1): 237–250.

63. Pan, Z., Ryu, H. and Baik, J. (2007) 'A Case study: CRM Adoption Success Factors Analysis and six sigma DMAIC Application'. *Proceedings of the 5th ACIS International Conference on Software Engineering Research, Management & Applications*, IEEE Computer Society, Washington, DC, USA.

64. Reponen, T. (2003) 'Information Technology-Enabled Global Customer Service'. IGI Publishing.

65. Richter, P. Cornford, J. and McLoughlin, I. (2004) 'The e-Citizen as talk, as text and as technology: CRM and e-Government'. *Electronic Journal of e-Government* 2 (3): 207–218.

66. Rigby, D., Reichheld, F. and Schefter, P. (2002) 'Avoid the Four Perils of CRM'. *Harvard Business Review* 80 (2): 101–109.

67. Rigby, D. K., Reichheld, F. F. and Schefter, P. (2002) 'Avoid the Four Perils of CRM'. *Harvard Business Review* January: 101–109.

68. Roberts-Witt, S. L. (2000) 'It's the Customer, Stupid'! *PC Magazine* June 27: 6–22.

69. Roha, T. H., Ahn, C. K. and Han, I. (2005) 'The priority factor model for customer relationship management system success'. *Expert Systems with Applications* 28: 641–654.

70. Salomann, H., Dous, M., Kolbe, L. and Brenner, W. (2005) 'Customer Relationship Management Survey, Status Quo and Future Challenges'. Institute of Information Management. University of St. Gallen.

71. Shang, S. and Seddon, P. (2003) 'Factors Affecting Net Benefits from Enterprise Systems'. Melbourne University Working Paper, cited in: Freeman, P. and Seddon, P. B. 'Benefits from CRM-Based Work Systems'. Available at: *http://is2.lse.ac.uk/asp/aspecis/20050017.pdf*

72. Siebel Systems (2004) 'Critical Success Factors for Phased CRM Implementations'. Siebel White Papers.

73. Simon, P. (ed.) (2010) 'The next wave of technologies: Opportunities in Chaos'. New York: John Wiley and Sons.

74. Soudagar, R., Iyer, V. and Hildebrand, V. (2011) 'The Customer Experience Edge: Technology and Techniques for Delivering an Enduring, Profitable and Positive Experience to Your Customers'. McGraw-Hill.

75. Starkey, M. and Woodcock, N. (2002) 'CRM Systems: Necessary, but not Sufficient. Reap the Benefits of Customer Management'. *Journal of Database Marketing* 9 (3): 267–275.

76. Stone, M., Woodcock, N. and Wilson, M. (1996) 'Managing the Change from Marketing Planning to Customer Relationship Management'. *Long Range Planning* 29 (5): 675–683.

77. Teece, D. J., Pisano, G. and Shuen, A. (1997) 'Dynamic Capabilities and Strategic Management'. *Strategic Management Journal* 18: 509–533.

78. Tembo, F. (2012) 'Citizen voice and state accountability: Towards theories of change that embrace contextual dynamics'. Overseas Development Institute, UK. Available at: *http://www.odi.org.uk/resources/docs/7557.pdf*

79. Thompson, A. A. and Strickland, A. J. (2001) 'Crafting and Executing Strategy' (12th Edition). Singapore: McGraw-Hill Irwin.

80. Thompson, B. (2002) 'What is CRM'? Available at: *www.crmguru.com*

81. Triki, A. and Zouaoui, F. (2011) 'Customer Knowledge Management Competencies Role in the CRM Implementation Project'. *Journal of Organizational Knowledge Management*: 1–11.

82. Tsiptsis, K. and Chorianopoulos, A. (2010) 'Data Mining Techniques in CRM: Inside Customer Segmentation'. Chichester, UK: John Wiley and Sons Ltd.

83. Wilson, H., Daniel, E. and McDonald, M. (2002) 'Factors for Success in Customer Relationship Management (CRM) Systems'. *Journal of Marketing Management* 18: 193–219.

84. Winer, R. (2001) 'A Framework for Customer Relationship Management'. California Management Review 43 (4): 89–105.

85. Wood, A. (2008) 'How to use segmentation for CRM success'. Available at: *http://www.utalkmarketing.com/Pages/Article.aspx?ArticleID=4219&Title =How_ to_use_segmentation_for_CRM_success*

86. Yarmoff, L. B. (2001) 'Empowering e-Government'. Available at: *http:// www.destinationcrm.com/Articles/CRM-News/Daily-News/Empowering-e-Government-45937.aspx*

87. Yu, L. (2001) 'Successful Customer-Relationship Management'. *MIT Sloan Management Review* 42 (4): 81–87.

Part 2
e-Government

e-Government strategies: the case of the United Arab Emirates

Abstract: This article provides an overview of e-government and its role in revolutionising existing governmental systems. It argues that for e-government initiatives to truly succeed, we need to develop public trust and confidence to promote diffusion and participation. The article relates this to the recently announced UAE e-Government Strategic Framework 2011–2013. The framework attempts to promote the electronic transformation of all government services within a period of three years. An important component of the strategic framework in question is the use of the existing national identity management infrastructure and the development of a government-owned federated identity management system to support Government-to-Citizen (G2C) e-government transactions and promote trust and confidence on the Internet.

Keywords: *e-Government, identity management, federated identity, identity card.*

1. Introduction

Information and Communications Technologies (ICT) have affected the ways in which people, governments, and businesses interact with each other. The rapid diffusion of the Internet, mobile telephony and broadband networks demonstrate how pervasive this technology has become. Today, ICT is considered as one of the fundamental building blocks of modern societies and digital economies (Castells, 2009; Varian et al., 2005).

Yet, the revolutionary pace in countries worldwide is dependent on the preparedness of several factors of both the social and political environments (Gauld and Goldfinch, 2006; Loader, 2009; OECD, 2009). New technologies have revealed their potential to threaten existing power settings and economic relationships (Beer, 2011; Nixon and Koutrakou, 2007). The numerous applications of ICT over the past few decades have shown its transformative potential and its usage as an important tool for organising political dissent in countries worldwide (Hirschfeld, 2012; Reddick, 2010; Serageldin, 2011).

From a government standpoint, e-government adoption is becoming an inevitable task. It deals with facilitating the operation of government and the distribution of governmental information and services. The ultimate goal of e-government is to be able to offer an increased portfolio of public services to citizens in an efficient and cost effective manner. Anticipated benefits include efficiency, improved services, better accessibility of public services, and more transparency and accountability (Atkinson and Castro, 2008), see also Figure 3.1.

The objective of this article is to examine some of the difficulties pertaining to the successful development and implementation of e-government programs. The aim is to be pragmatic and focus on the problematic areas from a practitioner's point of view, thus relating the identified concerns and mapping them to a case study drawn from the UAE e-government experience.

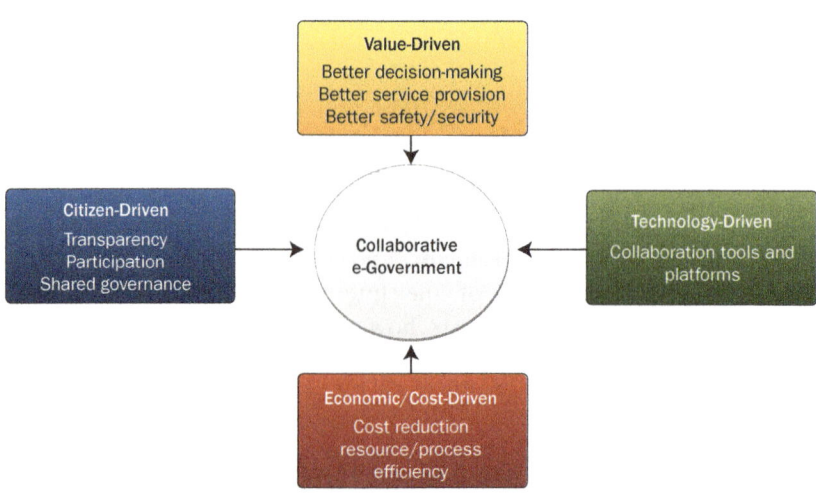

Figure 3.1 Primary drivers of e-government

The article is structured as follows: The first section provides a snapshot overview of the literature around the objectives and outcomes associated with e-government. It then briefly discusses the issue of trust and security in virtual networks and how it may encourage or inhibit public trust and confidence. The following section gives an overview of e-government in the Gulf Cooperation Council (GCC)* countries and some recent statistics about the spread of e-government. It then presents the case of the UAE e-Government Strategic Framework 2012–2014 and explains its primary objectives and components. Finally, it sheds some light on the UAE government's strategic initiative, the national identity management infrastructure and its federated identity management system explaining its potential role in supporting the e-government transformation and successful implementation of the government's strategy.

2. e-Government: the power of technology

e-Government in its simplest form is about the use of ICT to provide access to governmental information and deliver public services to citizens and business partners. However, practitioners have still not figured out how to exploit its full benefits. There is an equilibrium problem with e-government applications and limitations arising from the difficulty to tangibly justify the huge investments in ICT systems for the past decade and a half.

The average public expectations concerning governments' efforts are shaped according to the ability of the government to successfully improve citizens' quality of life. Governments need to ensure that their policies, regulations and systems enable citizen participation, and address the needs of improving the delivery of services. The service delivery lifecycle needs to be re-engineered and redesigned so as to meet citizen's expectations of enhanced social security and quality of life. Figure 3.2 depicts the role of government policy making in building a more citizen-centric and competitive government.

---

*GCC is the acronym for Gulf Cooperation Council, also referred to as the Cooperation Council for the Arab States of the Gulf (CCASG). It includes six countries, namely Bahrain, Kuwait, Oman, Qatar, Saudi Arabia, and the United Arab Emirates.

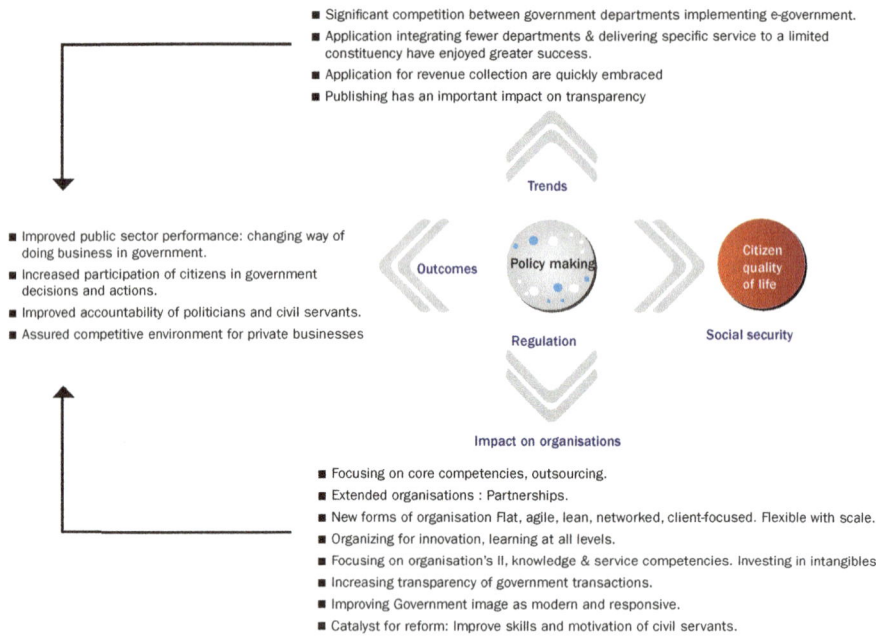

Significant competition between government departments implementing e-government.
Application integrating fewer departments & delivering specific service to a limited constituency have enjoyed greater success.
Application for revenue collection are quickly embraced
Publishing has an important impact on transparency

Trends

Improved public sector performance: changing way of doing business in government.
Increased participation of citizens in government decisions and actions.
Improved accountability of politicians and civil servants.
Assured competitive environment for private businesses

Outcomes   Policy making   Citizen quality of life

Regulation   Social security

Impact on organisations

Focusing on core competencies, outsourcing.
Extended organisations : Partnerships.
New forms of organisation Flat, agile, lean, networked, client-focused. Flexible with scale.
Organizing for innovation, learning at all levels.
Focusing on organisation's II, knowledge & service competencies. Investing in intangibles.
Increasing transparency of government transactions.
Improving Government image as modern and responsive.
Catalyst for reform: Improve skills and motivation of civil servants.

Figure 3.2   **Development of a new revolution in governments**

Government policies should enable governments to undertake radical organisational changes, that foster growth in services, reduce unnecessary costs and regulatory burdens on firms, strengthen education and training systems, encourage good management practices, foster innovation and new applications, foster market conditions and create a business environment that promotes a productive economy, and the list goes on.

Advocates of e-government point out the opportunities for citizens to play a greater role in public policy (Ambali, 2010; Bonina and Cordella, 2008; Navarra and Cornford, 2007; Torres et al., 2005). They also stress its potential to connect them, quickly and directly, to what their government has to offer – no queues, no waiting, service 24/7.

Cost-cutting is a major factor driving decisions to go online. Advanced e-government in our opinion has the potential to cut overheads by as much as 90 per cent through streamlined communications and integrated systems that offer higher levels of efficiency, effectiveness and convenience. That is to say, e-government initiatives can reduce administrative burdens, process time cycles and improve responsiveness. Besides, compared with the traditional over-the-counter services, online services can reduce substantial tangible costs as they, for example, do not need buildings, people, electricity, service desks and so on.

Indeed, ICT offers the potential for development and competition in the public sector specifically in areas of customer service and overall organisational excellence programs. Such competition not only helps lower the costs of government services through automation and computerisation but also increases pressure on firms to improve performance and change conservative attitudes.

The private sector has always challenged the public sector and acted as catalyst for better quality and for more effective budget utilisation (Suomi and Tähkäpää, 2002). Increased computerisation in the public sector is promoting new levels of balance between the two sectors (Das et al., 2010). Government agencies and public sector agencies in particular are paying greater attention to core capabilities, and outsourcing other support functions to be delivered by the private sector (Suomi and Tähkäpää, 2002). ICT in this regard has played a central role in helping governments to achieve remarkable productivity gains (EIU, 2004).

On the other hand, and despite high spending and the widespread adoption of sophisticated ICT infrastructure, many other countries continue to lag behind on key measures of economic growth and productivity (ibid). Government investment in ICT to date has been very narrowly focused on administrative rationalisation, cost-cutting, and service reform without giving attention to creating public oriented systems that promote and encourage citizen participation (Longford, 2002).

The major deficiency in such efforts is that they have been thought of and executed from a 'government mindset' rather than being based on public needs and expectations. Such a narrow view of e-government calls for reported ICT achievements to be regarded with a skeptical eye (Longford, 2002). Unless measures are taken to address other aspects of society and governance, e-government alone may produce little if any net gain in leveraging ICT to rationalise and restructure administrative systems and service delivery systems (ibid).

Other researchers recommend that governments adopt a new approach that embeds a transformation in the logic underpinning the design and evaluation of public sector organisations (Lane, 2000). This is envisaged to have considerable implications for improving the services delivered by public administration, and serious consequences for the public value associated with the services delivered (Bonina and Cordella, 2008).

In Arab countries, e-government is now viewed as the path to develop a more sustainable new economy. It is also considered as playing a vital role in managing and directing the process of change and reform that will boost public confidence. However, building trust in e-government is not a simple issue. The relevant literature shows that there are overwhelming concerns about the potential of digital networks to negatively affect public privacy and security (Conklin and Whiet, 2006; McLeod and Pippin, 2009; Nikkhahan et al., 2009; Palanisamy and Mukerji, 2012; Yee et al., 2005). The next section discusses this in more detail.

3. Trust and confidence

Trust is probably one of the most important aspects in the implementation of e-government strategies. In order for e-government to achieve its ambitious objectives to develop and deliver high quality and integrated public services, citizens need to trust the virtual environment. Without trust, citizens will not participate in the e-government process.

A review of the literature and empirical studies on e-government identifies the criteria for its adoption from both a citizen's and government's perspective, which highlights trust and security as major factors (Al-Khouri, 2012a; Tassabehji and Elliman, 2006). Empirical evidence shows that the level of trust is simply not a gradual process that happens over time (Berg et al., 1995; Kramer, 1999), but rather a cumulative process. There are several overlapping and consistent factors that have the potential to impact the building of trust. These are classified in two major clusters; pre-interactional and interactional factors, as depicted in Table 3.1 (Colesca, 2009).

For the successful adoption of e-government services, citizens must have the intention to 'engage in e-government' which encompasses the intentions to receive and provide information through online channels (Warkentin et al., 2002). With the increasing reach of digital communication tools and connectivity, governments' interactions with their citizens over virtual networks are becoming more popular. Citizens have come to expect and demand governmental services matching private-sector services in every aspect of quality, quantity, and availability.

In fact, such expectations put more pressure on governments to develop quality services and delivery systems that are efficient and effective. However, the complexity arises from the fact that a citizen plays multiple roles while interacting with the government. Single role-based identities are decreasingly relevant in existing government transactions. This makes it imperative for governments to acquire citizen-centric qualities that provide services and resources tailored to the

Table 3.1 Factors that impact the building of trust

Pre-interactional factors	
Individual citizen/ consumer behavioural attributes	Subjective norms, individual demographics, culture, past experiences, propensity to trust, benevolence, credibility, competency, fairness, honesty, integrity, openness, general intention to trust and use of e-services.
Institutional attributes	Organisational reputation, accreditation, innovativeness, general perceived trustworthiness of the organisation.
Technology	Hardware and software that deliver security and effectiveness such as interface design, public key encryption, integrity.
Interactional factors	
Product/service attributes	Reliability, availability, quality, and usability.
Transactional delivery and fulfilment of services	Usability, security, accuracy, privacy, interactivity, quality.
Information content attributes	Completeness, accuracy, currency, quality.

actual service and resource needs of the users, including citizens, residents, government employees, business partners, and so on.

The next section provides a snapshot of e-government in GCC countries who have been recognised globally for their efforts in e-transformation and e-readiness.

4. e-Government in GCC countries

The latest United Nations Development Programme (UNDP) report on e-government shows a high level of preparedness in Middle Eastern countries, well above the world average in terms of e-government adoption and readiness to interact proactively with citizens. Internet usage in the Middle East is reported to be 35.6 per cent, compared to 32.6 per cent worldwide (UNDP, 2012). See also Table 3.2.

Table 3.2   Internet users in the Middle East and the rest of the world

Internet Users in the Middle East and in the world						
Middle East Region	Population (2011 Est.)	Pop.% of World	Internet Users 31 Dec 2011	Population % (Penetration)	Users % World	Facebook 03 2012
Total Middle East	216,258,843	3.1%	77.020,995	35.6%	3.4%	20,247,900
Rest of the World	6,713,796,311	96.9%	2,190,212,747	32.6%	96.6%	815,277,380
World Total	6,930,055,154	100%	2,267,233,742	32.7%	100%	835,525,280

Source: http://www.internetworldstats.com

Representing a total of 77 million Internet users, Middle Eastern citizens are classified as heavy users of electronic social networks with high dependence on digital communications. The United Arab Emirates have the highest Internet penetration with nearly 70 per cent of the population, followed closely by Qatar, Bahrain, Oman, Kuwait, Palestine and KSA. See also Figure 3.3.

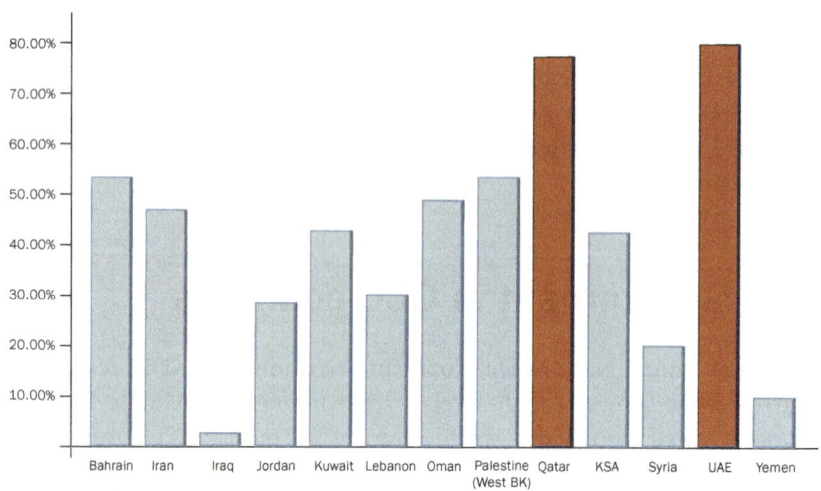

Figure 3.3   Percentage of the population with Internet access in Middle Eastern countries

Source: http://www.internetworldstats.com

Overall, GCC countries have maintained leadership in e-government readiness among Arab nations. They have taken serious steps to support the diffusion of e-government in their societies (Al-Khouri and Bachlaghem, 2011; Al-Khouri and Bal, 2007). Several UNDP reports confirmed that the growing efforts of GCC governments to promote digital transformation and literacy have helped further enhance the region's collective ranking in the UN e-Government Readiness Surveys (UNDP, 2010; UNDP 2012). These reports indicated that GCC countries played various roles for e-government in addressing the global financial crisis.

Governments of the GCC countries are considered to be in intense competition with each other to develop a new knowledge-based economy, away from the current dependence on oil, and to make their products and services competitive on a global scale (Awan, 2003). GCC countries are proceeding at a rapid pace to use more service oriented and citizen-centric operating models. This rapid reform is bringing a paradigm shift in the way citizens in the GCC are interacting with their governments. There are serious efforts in these countries to develop electronic operating environments, with advanced capabilities to build the right conditions for the e-citizen concept to evolve.

The next section provides an overview of the e-government strategy of one of the GCC countries, namely the UAE government's strategic framework that aims to electronically transform all public services through a two-year action plan.

5. UAE e-government strategic framework 2012–2014

Although local initiatives in the UAE began earlier, the federal e-government program started in 2001. One of the early e-services offered at a federal level was the electronic card known as the e-Dirham in 2001, which was issued to collect government services fees (Figure 3.4). Today, the UAE is considered to have one of the most advanced and world-class information and communication technology infrastructures.

The UAE is considered among the highest investing governments in adopting and implementing progressive ICT in its government and private sectors. The UAE has made a remarkable global achievement in the field of e-government according to the UN e-Government Survey 2012, which focuses on the role of e-government in sustainable development. The UAE achieved 28th rank overall according to the

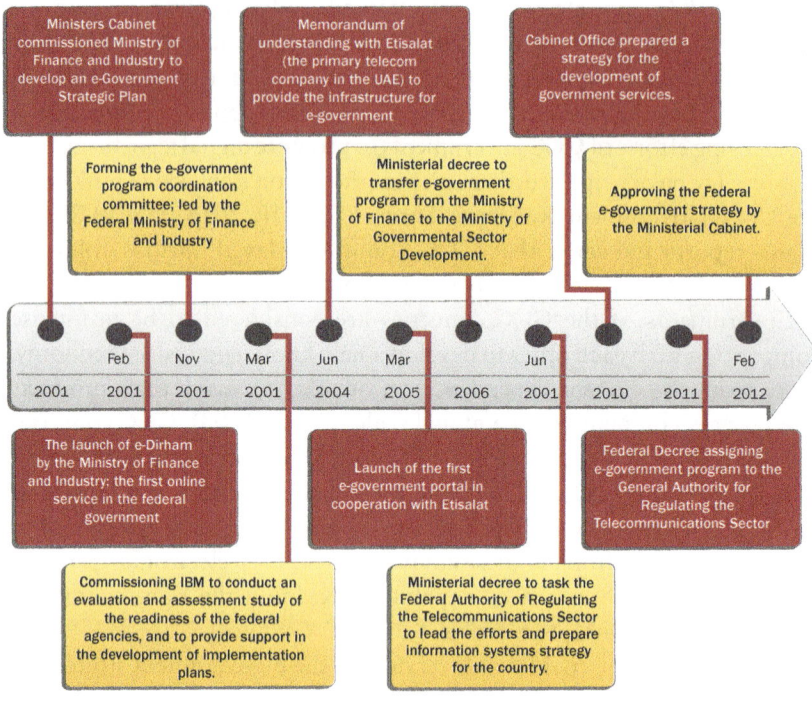

Figure 3.4 UAE federal e-government evolution

survey compared with 49th in the 2010 Survey. It came seventh on an online service index compared to 99th in 2010, and sixth on an e-participation index compared to 86th in 2010 (Figure 3.5).

The UAE has recently announced a revised e-Government Transformation Strategic Framework. This framework comprises numerous strategic initiatives at a federal level to transform all government services and make them available electronically through various channels. The following section will provide an overview of this strategy.

5.1 The UAE federal e-government strategic framework

The United Arab Emirates has developed a federal e-government strategic framework for 2012-2014 that charts out the initiatives and courses of action the government intends to take over a period of three years. The framework is aimed to contribute to:

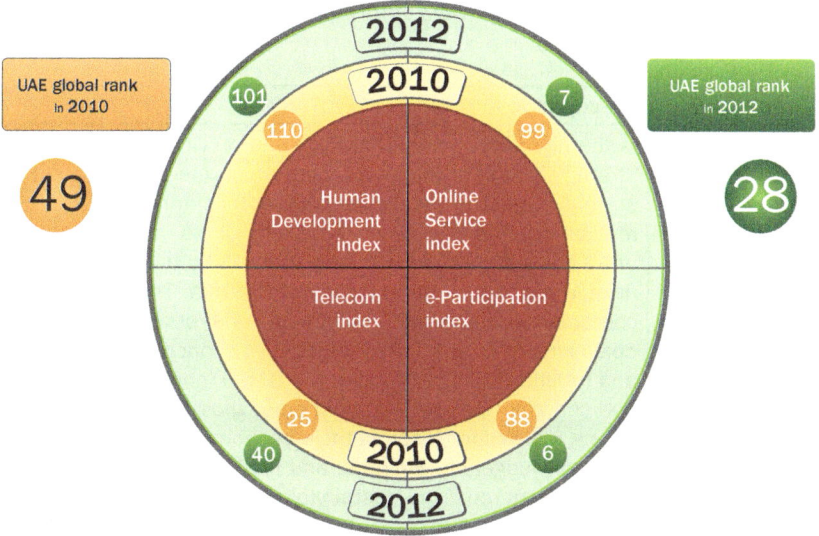

Figure 3.5   The UAE in the 2012 UN e-Government Survey

1. UAE Vision 2021, which drives the UAE to be one of the best countries in the world, see also Table 3.3.
2. UAE Government Strategy 2011–2013, that aims at putting citizens first and developing an accountable and innovative government.

The framework also makes reference to some of the existing federal strategies to ensure alignment with government strategic plans. See also Table 3.3.

There is considerable leadership confidence that successful implementation of the federal e-government strategy 2012–2014 will help to improve the UAE's global competitiveness and enhance the UAE's e-transformation. This is described clearly in the vision and mission statements developed as part of the strategy and as depicted in Figure 3.6.

As shown in the above diagram, the government adopted a seven-stage strategic development process. It included benchmarks with some international e-government practices and implementations, such as Canada, USA, Southern Europe, Singapore, the European Union, and GCC countries. The outcome of this exercise was the definition and prioritising of the initiatives and the primary focus areas. The development approach took into account three primary dimensions of e-services, e-readiness, and the ICT environment (Figure 3.7).

| Table 3.3 | The seven primary references in the UAE e-government strategy |

References	Description
UAE Vision 2021	Is the highest reference strategy and provides the strategic vision of the country, for which the e-government strategy needs to be aligned with, and contribute to its realisation. The UAE vision 2021 envisages development of a knowledge-based economy that will be diverse and flexible led by skilled professional Emiratis. The vision contains four important components with detailed objectives related to national identity, economy, education and health. It seeks to make the UAE a land of ambitious and confident people who hold on to their heritage; a strong federation; a competitive economy led by creative and knowledgeable Emiratis; and finally a high quality of life in a generous and sustainable environment. *http://www.vision2021.ae/*
UAE Strategy 2011–2013	Provides a phased plan for the Federal Government to progress towards the UAE Vision 2021. *http://uaecabinet.ae/English/Documents/PMO%20StrategyDocEngFinV2.pdf*
UAE Government ICT Strategy	Government strategy to regulate the telecommunications sector. It represents the basis on which the e-government strategy was developed, as it defines and details the three dimensions of service, environment and readiness.
Services Development Strategy	Provides an analysis of the current state of federal government services, as well as detailed guidelines on how to develop them. It also includes many of the strategic initiatives that fall under the e-government programme.
Federal government budget	Alignment of e-government budget with the federal budget. *http://www.mof.gov.ae/En/Budget/Pages/ZEROBudgeting.aspx*
Current Situation Analysis	Covers three dimensions (environment, readiness and services), and contributes to the identification of gaps and opportunities that can be addressed through the objectives and specific initiatives in the e-government strategy. *http://www.emiratesegov.ae*
Benchmarking	Comparisons of best practices in the field of e-government to support the development of the new strategy and define its primary objectives and initiatives.

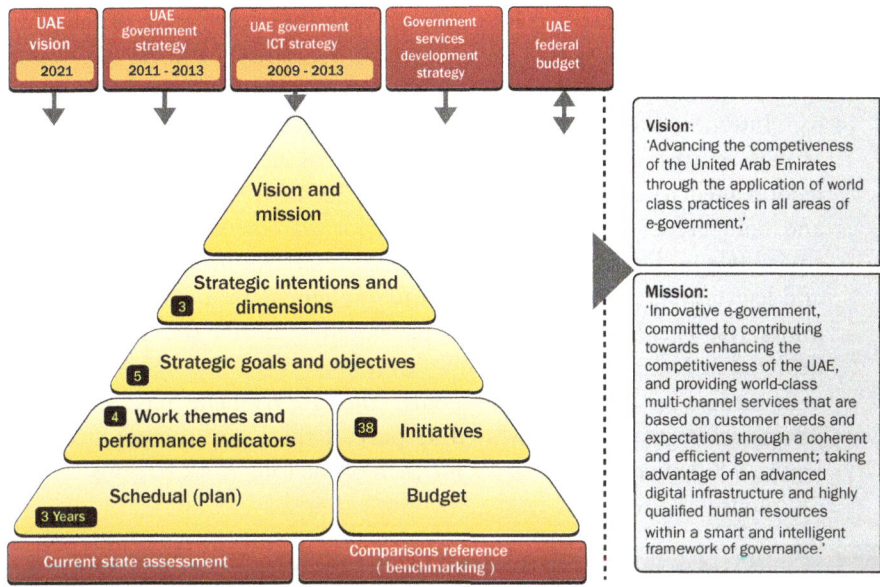

Figure 3.6 UAE e-government development methodology

Source: http://www.emiratesegov.ae

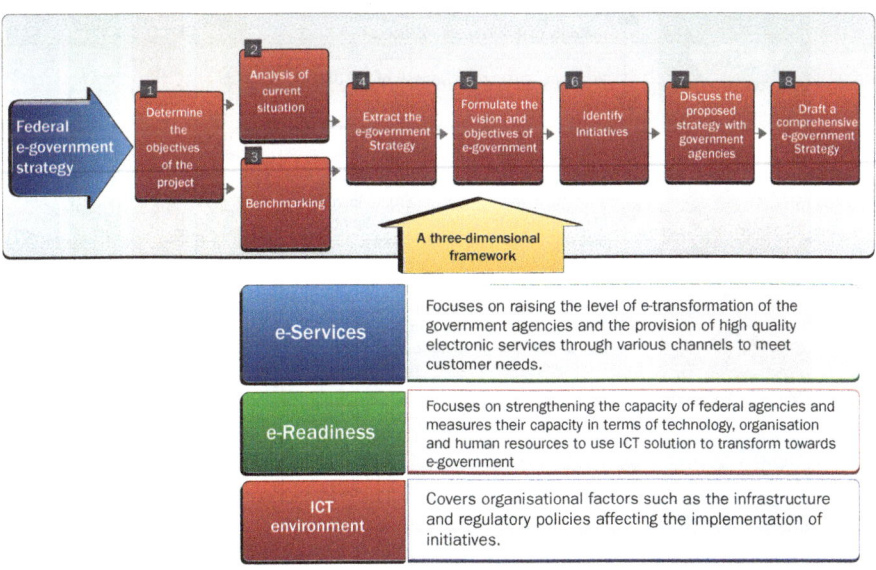

Figure 3.7 Strategic development plan

Source: http://www.emiratesegov.ae

The e-service dimension is concerned with the acceleration of the pace of e-transformation within government organisations and the provision of high quality electronic services through innovative delivery channels; e.g., Internet, fixed and mobile phones and kiosks, besides the traditional service centres. e-Readiness focuses on strengthening the capacities of federal agencies in terms of ICT, organisation structures, HR capabilities and competencies, and their readiness for e-transformation. The ICT environment dimension covers organisational factors such as policies and legislations needed to support the implementation of e-government initiatives. This has resulted in the development of five strategic goals as depicted in Figure 3.8.

In order to achieve these goals, the government has identified 38 initiatives to be implemented as part of the e-government strategy. Figure 3.9 depicts the initiatives for each of the four work streams. These 38 initiatives cover four vital e-government areas.

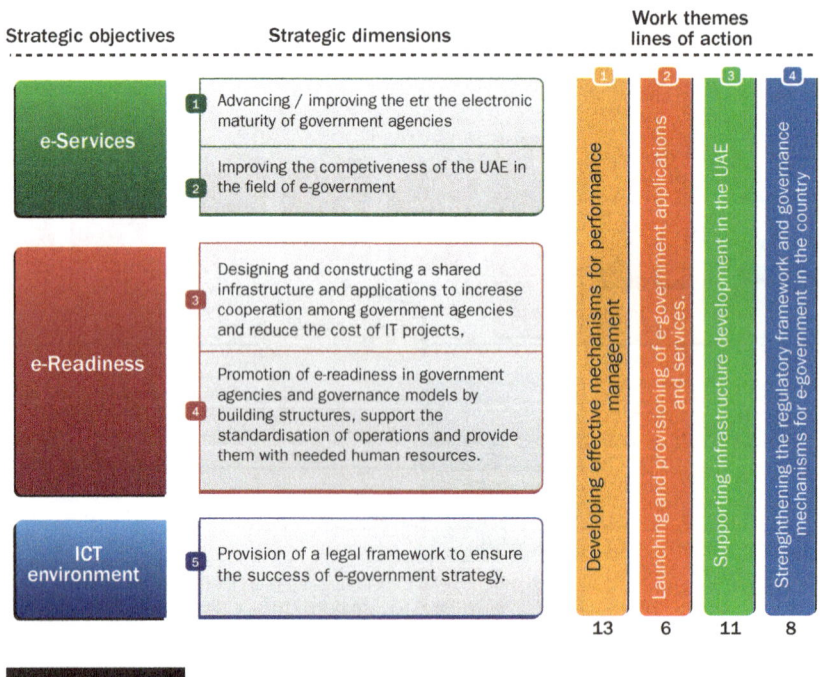

Figure 3.8 Strategic intentions, goals, and work themes

Strategic Goals

* advancing/raising the e-transformation levels in government services
* improving the competitiveness of UAE in the field of e-government.
* Establishing advanced infrastructure and promote cooperation between government agencies.
* promotion of e-readiness in government agencies and governance models
* Provision of legislative and legal to ensure the success of e-government strategy.

Work Themes

First axis Strengthening the regulatory framework and governance mechanisms for e-government in the country.	1. Development of a legal framework for e-government 2. Development of e-government security policies 3. Development of e-government policies 4. Development of a management system to measure & monitor e-transformation levels of government agencies. 5. Development of e-transformation strategy for government services. 6. Development of quality standards for e-services. 7. The launch of an electronic platform to support the development of e-Content in government agencies. 8. Design of the institutional structure of the UAE government (AE). 9. Development of Cloud computing strategy 10. Development of the electronic maturity model. 11. Development of ICT skills and competencies strategy for federal government agencies. 12. Development of ICT Structure for federal government agencies. 13. Development of e-government environment-friendly, green ICT strategy and policies.
Second axis Infrastructure support of information systems in the UAE	1. Build and operate the FedNet information network for the UAE government. 2. Development of data sharing model between government agencies 3. Building and managing the data center and disaster recovery 4. Launch and operation of customer relationship management centre 5. Development of an integration (interoperability) model for federal government systems. 6. Provide e-mail services to the staff of the federal government agencies.
Third axis Launching and availing e-government applications and services	1. Development and management of the official portal of the government of the UAE. 2. Update standards for federal government agencies websites and conduct annual evaluation. 3. Launch and management of electronic encyclopedia of the UAE. 4. Launch and operation of e-government service via mobile phones and devices. 5. The launch of open-to-all e-government program 6. The launch of UAE e-Government Award. 7. Specialized electronic portals - Phase I. 8. Construction of Cloud computing for UAE government. 9. Launch and operation of knowledge management system. 10. Development of e-Service Service Level Standards. 11. Build a model to calculate the return on investments for e-government projects.
Fourth axis Development of effective mechanisms for performance management	1. Measuring customer satisfaction on e-services 2. Measuring the quality of e-government services 3. Media Campaign to enhance the level of awareness and use of public access e-services and the different channels. 4. Develop and implement an action plan to improve the UAE ratings on the United Nations e-Government Index. 5. Systematic documentation and marketing of e-government success stories. 6. The development of a dissemination framework of successful practices in e-government. 7. The development of business continuity governmental policies. 8. Measuring the e-maturity of government agencies.

Figure 3.9   UAE e-government 2011–2014 initiatives

Source: http://www.emiratesegov.ae

1. Strengthening the regulatory framework and governance mechanisms for e-government. This is related to the legal and regulatory environment governing acquisition and use of information systems in

government agencies, e-government services, and a high level plan for the overall development of the public sector. Regulations and laws are considered primary enablers to support e-government and ensure security, reliability and data privacy. As such, this area also includes the development of a strong governance structure to facilitate communication between the different stakeholders and attempts to capture their needs and turn them into electronic service systems.

2. Infrastructure support of information systems in the UAE. This theme deals with creating a solid infrastructure for information systems to enable the delivery of world-class e-government services. It also focuses on aspects such as facilitation of exchange and sharing of data between government agencies.

3. Launching and providing e-government applications and services. This theme focuses on a set of applications and services to be provided to government agencies to support them to provide e-government services effectively and efficiently.

4. Development of effective mechanisms for performance management. This theme focuses on improving overall effectiveness and actual levels of performance of departments of information technology within government agencies. It also deals with developing automated tools and reports to monitor performance indicators and overall performance management.

The government identified 20 strategic performance indicators across all five strategic objectives to measure the implementation success of the strategy. Figure 3.10 shows eight of these key performance indicators (KPIs).

The government also developed an operating model that will be used to measure progress based on two variables: citizen centricity and efficiency and effectiveness factors associated with initiatives and projects. The model consists of six elements, as depicted in Figure 3.11. Each of these elements is managed through a separate and dedicated set of project portfolios. The most important element in the model is the construction of necessary security measures to develop trust and confidence levels between the service providers and the beneficiaries.

One of the key programs launched by the UAE to build trust and security in its e-government plan is the national identity management infrastructure programme. There is a high level of interdependence between these two initiatives. As part of the program, the UAE issues a smart card with digital identities for all of its population, which is estimated at around nine million people. The next section will elaborate further on the objective of this program.

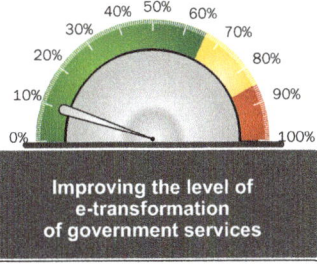

Improving the level of e-transformation of government services

1. Percentage of e-transformation of government services.
2. Percentage of (individual) use of e-services.
3. Percentage of (business enterprises) use of e-services.
4. Percentage of customer satisfaction (individuals) on e-Government services.
5. Percentage of customer satisfaction (business enterprises) on e-government services.
6. Percentage of compliance of government agencies websites to e-government standards.
7. Percentage of compliance of government agencies e-services to e-government standards.

Development of an advanced infrastructure to promote cooperation government agencies

1. Percentage of available cloud services.
2. Percentage of government agencies using cloud services.
3. value/cost savings from the use of cloud services.
4. Percentage of completion of Enterprise Architecture in government agencies.
5. Level of integration between government agencies.
6. Percentage of compliance of government agencies with the government ICT standards.

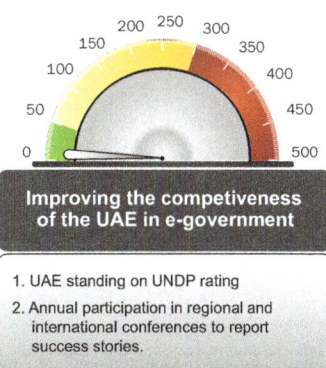

Improving the competiveness of the UAE in e-government

1. UAE standing on UNDP rating
2. Annual participation in regional and international conferences to report success stories.

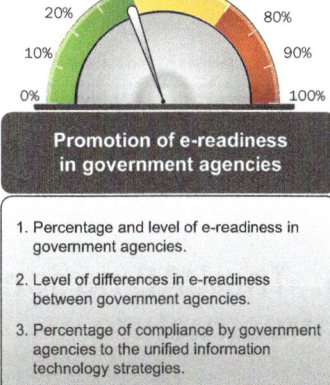

Promotion of e-readiness in government agencies

1. Percentage and level of e-readiness in government agencies.
2. Level of differences in e-readiness between government agencies.
3. Percentage of compliance by government agencies to the unified information technology strategies.

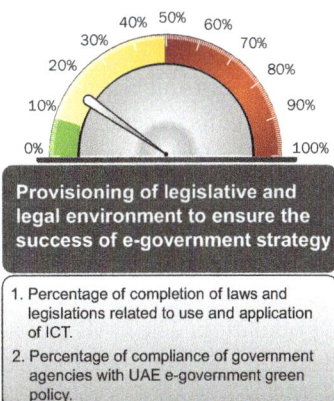

Provisioning of legislative and legal environment to ensure the success of e-government strategy

1. Percentage of completion of laws and legislations related to use and application of ICT.
2. Percentage of compliance of government agencies with UAE e-government green policy.

Figure 3.10   Examples of UAE e-government KPIs 2011–2014

Source: http://www.emiratesegov.ae

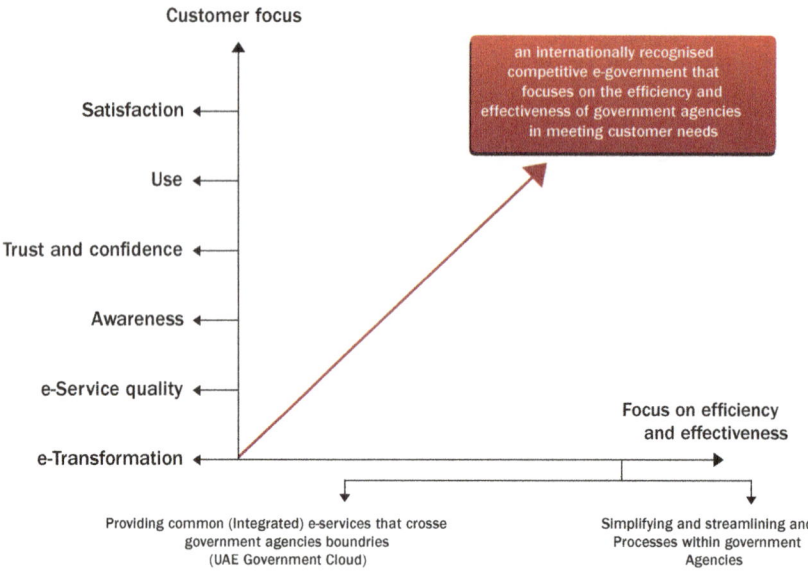

Customer focus

an internationally recognised competitive e-government that focuses on the efficiency and effectiveness of government agencies in meeting customer needs

Satisfaction

Use

Trust and confidence

Awareness

e-Service quality

e-Transformation

Focus on efficiency and effectiveness

Providing common (Integrated) e-services that crosse government agencies boundries (UAE Government Cloud)

Simplifying and streamlining and Processes within government Agencies

Figure 3.11   **UAE e-government strategic operating model**

Source: http://www.emiratesegov.ae

6. The UAE national identity management infrastructure

The UAE national identity management infrastructure is a strategic initiative to enhance homeland security and develop a federated identity management system enabling secure e-government transactions (Al-Khouri, 2012b). A federated identity is the means of linking a person's electronic identity and attributes stored across multiple distinct identity management systems (Madsen, 2005). Such systems would allow individuals to use the same user name, password or other personal identification to sign in to the networks of more than one enterprise in order to conduct transactions (Bertino and Takahashi, 2011; Roebuck, 2011; Windley, 2005).

As part of the program, the UAE issues smart identity cards for all of its population. The UAE national identity card is one of the world's most advanced and secure smart cards. The card is provided with identification parameters stored securely in a smart chip. It thus enables the establishment of a person's identity on-site (physically) and remotely (virtually), enabling secure and trusted transactions. The multi-factor authentication which provides both match-on-card and match-off-card features, facilitates validation, verification and authentication of any given identity. The card holder can then access all identity-based services as shown in Figure 3.12.

The UAE ID card capabilities of on-site and remote identification and authentication are available to be used across different applications enabling various forms of electronic transactions e.g., G2C, B2C and so on. These are facilitated by PIN verification, biometric authentication (match-on-card and match-off-card features), and digital signatures (Figure 3.13).

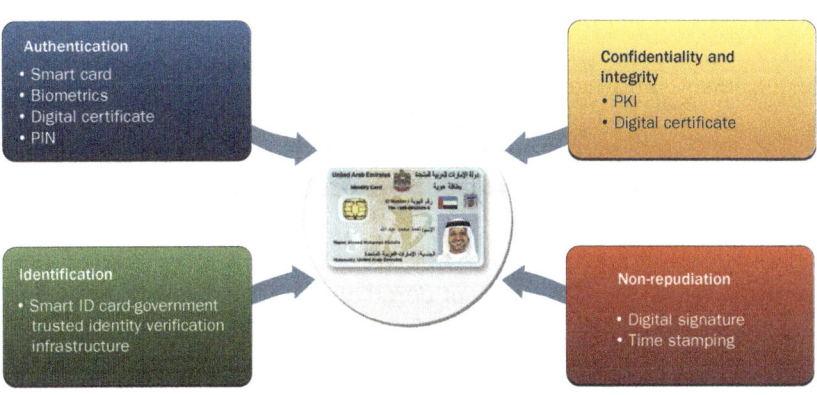

Figure 3.12 The national ID card; key enabler for UAE e-government

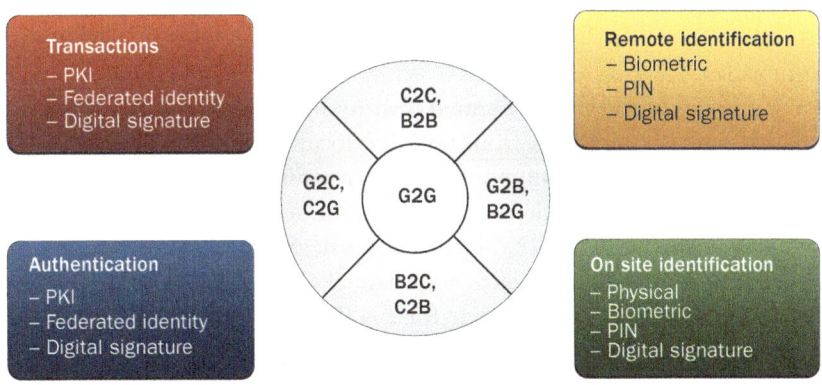

Figure 3.13 Enabling secure e-government transactions through smart identity cards

The UAE national identity management system eliminates the need to maintain distinct user credentials in separate systems. In an e-government context, this should result in greatly simplified administration and streamlined access to resources.

Government agencies in the UAE federated identity management (FIM) system will depend on the National Identity Validation Gateway to authenticate their respective users and vouch for their access to services. Agencies will be able to share applications without needing to adopt the same technologies for directory services, security and authentication. This is enabled by the active directory services of the FIM that allow government agencies to recognise their users through a single identity (Figure 3.14).

The UAE is currently taking rapid steps in integrating its identity management infrastructure and its smart card capabilities in various public sector systems and applications. Some of the current deployments for card usage include the e-Gate service at the airports that allows card holders to pass through immigration control using biometric authentication.

In addition, citizens in Abu Dhabi, for example, have the ability to log on to the online local government portal and avail themselves of various e-services and utility payments. Some additional services provided through the Abu Dhabi portal include viewing and modifying details of one's personal traffic profile with the Abu Dhabi Police, such as address, licence plate and so on.

There is increasing motivation in the UAE public sector to rely on the new identity card to provide its services. It is expected that all e-government services will eventually require registering for the UAE identity card and PIN to access online government services. Integration of the national identity card is ongoing in all federal and local authorities.

The design of the UAE federated identity management system ensures reliable and secure access from multiple locations, and hence provides advanced mobility. This supports the vision set in the UAE e-government strategic framework to deliver public sector services through different channels; whether it is the Internet, kiosk machines, mobile phone applications or any other electronic channel. The UAE national identity card is viewed as the cornerstone for enabling successful deployment of e-government and e-service strategy in the country.

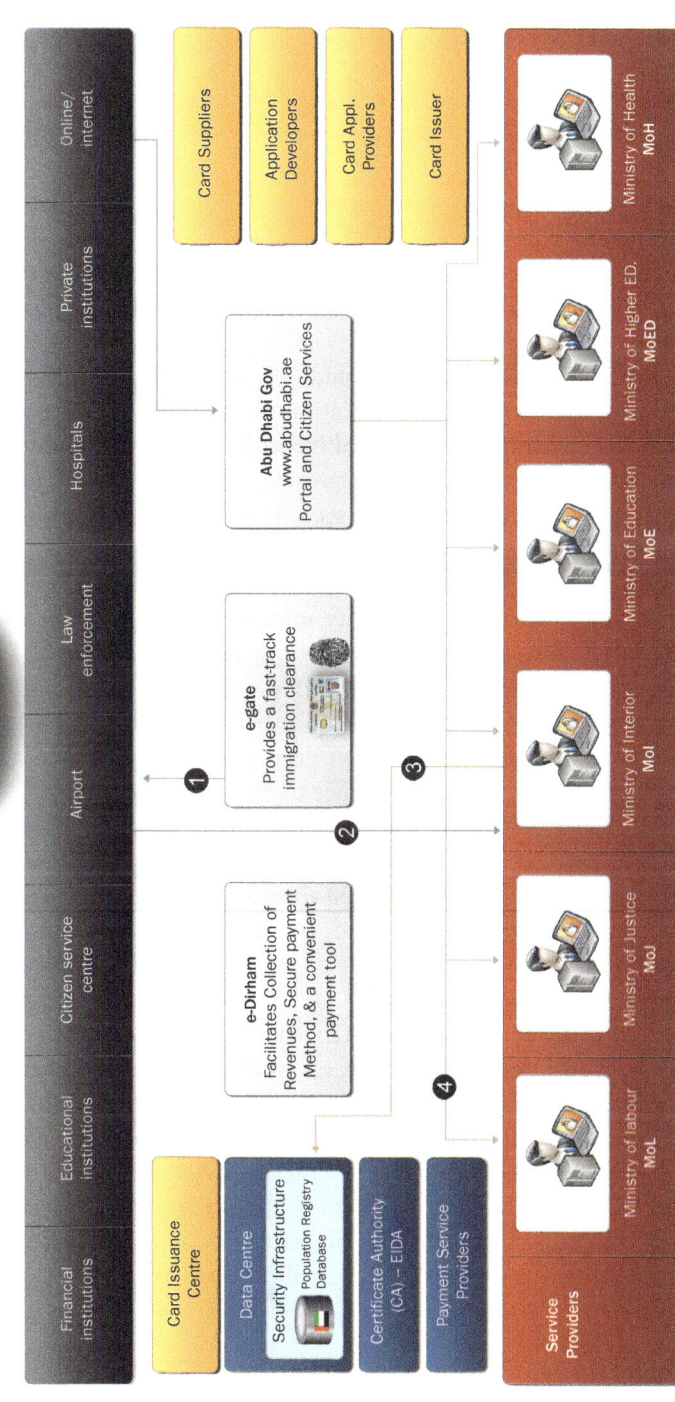

Figure 3.14 Federated identity management system

7. Conclusion

In an era of increasing digital communications and connectivity, governments are paying more attention to interaction with their citizens within the virtual world (Bwalya, 2012; Reddick, 2010b). While making such attempts, governments are realising that conventional physical trust mechanisms are now insufficient and that there is a clear need to develop new capabilities to identify electronic identities (Andress and Winterfield, 2011; Basin et al., 2011; Howard and Prince, 2011; Sheldon and Vishik, 2011).

The UAE government decided, as part of its national development strategy, to own the identification process itself and provide secure, unique and tamperproof digital identities to its population. This kind of identity management system owned by the national government is envisaged to offer improved security, and gain higher levels of trust, confidence and encourage participation.

The federated identity management system, which is a fundamental component of the UAE identity infrastructure, is foreseen to eliminate the need to replicate databases of users' credentials for separate applications and systems. It also paves the way to use a common framework to share information between trusted partners, where government agencies would not need to establish separate relationships and procedures with one another to conduct transactions.

The UAE e-government initiatives will be more successful when citizens are able to transcend physical borders to carry out their transactions. A citizen should be able to use his/her national identity card to conduct e-government and e-commerce transactions on websites verified and validated by a single identity validation service. This should be the future aspiration.

To the extent that the UAE federated identity allows government agencies to offer controlled access to data or other resources, it has the potential to enable new levels of collaboration between the different agencies. Identity management can support process re-engineering for extending access to valuable resources, using multi-factor authentication mechanisms, while the integration of systems across governmental and private sector spheres further broadens the opportunities for supporting e-government and e-commerce applications.

References

1. Al-Khouri, A. M. and Bachlaghem, M. (2011) 'Towards Federated e-Identity Management across the GCC: A Solution's Framework'. *Global Journal of Strategies & Governance* 4 (1): 30–49.
2. Al-Khouri, A. M. and Bal, J. (2007) 'Electronic Government in the GCC Countries'. *International Journal Of Social Sciences* 1 (2): 83–98.
3. Al-Khouri, A. M. (2012a) 'Emerging Markets and Digital Economy: Building Trust in The Virtual World'. *International Journal of Innovation in the Digital Economy* 3 (2): 57–69.
4. Al-Khouri, A. M. (2012b) 'PKI in Government Digital Identity Management Systems'. *The European Journal of e-Practice* 14: 4–21.
5. Ambali, A. (2010) 'E-government in Public Sector: Policy Implications and Recommendations for Policy-makers'. *Research Journal of International Studies* 17. Available at: *http://www.eurojournals.com/RJIS_17_10.pdf*
6. Andress, J. and Winterfield, S. (2011) 'Cyber Warfare: Techniques, Tactics, and Tools for Security Practitioners'. Waltham, MA: Syngress.
7. Atkinson, R. D. and Castro, D. D. (2008) 'Digital Quality of Life: Understanding the Personal and Social Benefits of the Information Technology Revolution'. *Information Technology and Innovation Foundation* March: 137–145. Available at: *www.itif.org/files/DQOL.pdf*
8. Awan, M. (2003) 'e-Government: Assessment of GCC (Gulf Co-operating Council) Countries and Services Provided'. *Lecture Notes in Computer Science* 2739: 500–503.
9. Basin, D., Clavel, M. and Egea, M. (2011) 'A Decade of Model-Driven Security'. In Breu, R., Crampton, J. and Lobo, J. (eds), *SACMAT, Proceedings of the 16th ACM Symposium on Access Control Models and Technologies:* 1–10. Innsbruck, Austria, June 15–17.
10. Beer, W. (2011) 'Cybercrime: Protecting against the Growing Threat, Global Economic Crime Survey'. Available at: *http://www.pwc.com/en_GX/gx/economic-crime-survey/assets/GECS_GLOBAL_REPORT.pdf*
11. Berg, J., Dickhaut, J. and McCabe, K. (1995) 'Trust, Reciprocity and Social History'. *Games and Economic Behavior* 10 (1): 122–142.
12. Bertino, E. and Takahashi, K. (2011) 'Identity management: Concepts, technology, and systems'. Boston, MA: Artech House.
13. Bonina, C. M. and Cordella, A. (2008) 'The New Public Management, e-Government and the Notion of "Public Value". 'Lessons from Mexico'. *Proceedings of SIG GlobDev's First Annual Workshop*. Paris, France December 13th. Available at: *http://www.globdev.org/files/24-Paper-Bonina-The%20 New%20Public%20Mgt-Revised.pdf*
14. Bwalya, K. J. (2012) 'E-Government in Emerging Economies: Adoption, E-Participation, and Legal Frameworks'. Hershey, PA: IGI Global Publishing.
15. Cairncross, F. (1997) 'The Death of Distance: How the Communications Revolution will Change our Lives'. London: Orion Business Books.

16. Castells, M. (2009) 'The Rise of the Network Society: The Information Age'. *Economy, Society, and Culture* 1 (2).

17. Colesca, S. E. (2009) 'Understanding Trust in e-Government'. *Inzinerine Ekonomika – Engineering Economics* (3): 7–15.

18. Conklin, A. and Whiet, G. B. (2006) 'E-Government and Cyber Security: The Role of Cyber Security Exercises'. *Proceedings of the 39th Annual Hawaii Imitational on System Sciences*: 79b. Hawaii.

19. Das, S., Krishna, K., Lychagin, S. and Somanathan, R. (2010) 'Back on the Rails: Competition and Productivity in State-owned Industry'. *NBER Working Paper* 15976. Available at: *http://www.nber.org/papers/w15976.pdf*

20. EIU (2004) 'Reaping the benefits of ICT Europe's productivity challenge'. A report from the Economist Intelligence Unit sponsored by Microsoft. Available at: *http://graphics.eiu.com/files/ad_pdfs/MICROSOFT_FINAL.pdf*

21. Gauld, R. and Goldfinch, S. (2006) 'Dangerous enthusiasms: e-Government, computer failure and information systems development'. Dunedin, New Zealand: Otago University Press.

22. Hirschfeld, B. (2012) 'Global Thesis Update: Technology and the Arab Spring'. Available at: *http://worldperspectivesprogram.wordpress.com/2012/04/12/global-thesis-update-technology-and-the-arab-spring/*

23. Howard, D. and Prince, K. (2011) 'Security 2020: Reduce security risks this decade'. Hoboken, NJ: John Wiley and Sons.

24. International Data Corporation (IDC) (2011) *www.idc.com*

25. Kramer, R. (1999). 'Trust and Distrust in Organisations: Emerging Perspectives, Enduring Questions'. *Annual Review of Psychology* 50: 569–598.

26. Lane, J. E. (2000) 'New Public Management: An Introduction'. London: Routledge.

27. Loader, B. (2009) 'Beyond e-Government'. London: Routledge.

28. Longford, G. (2002) 'Rethinking e-Government: Dilemmas of Public Service, Citizenship and Democracy in the Digital Age'. *The Public Sector Innovation Journal*. Available at: *http://www.innovation.cc/news/innovation-conference/longford.pdf*

29. Madsen, P. (2005) 'Liberty Alliance Project White Paper: Liberty ID-WSF People Service – Federated Social Identity'. Available at: *http://www.projectliberty.org/liberty/content/download/387/2720/file/Liberty_Federated_Social_Identity.pdf*

30. McLeod, A. J. and Pippin, S. E. (2009) 'Security and Privacy Trust in e-Government: Understanding System and Relationship Trust Antecedents'. *Proceedings of the 42nd Annual Hawaii International Conference on System Sciences HICSS*.

31. Navarra, D. D. and Cornford, T. (2007) 'The State, Democracy and the Limits of New Public Management: Exploring Alternative Models of e-Government. Information Systems Group'. The London School of Economics and Political Science. Available at: *http://is2.lse.ac.uk/wp/pdf/wp155.pdf*

32. Nikkhahan, B., Aghdam, A. J. and Sohrabi, S. (2009) 'e-Government Security: A Honeynet Approach'. *International Journal of Advanced Science and Technology* 5: 75–84.
33. Nixon, P. G. and Koutrakou, V. N. (eds.) (2007) 'e-Government in Europe: Re-booting the State'. London: Routledge.
34. OECD (2009) 'Rethinking e-Government Services: User-Centred Approaches'. Organization for Economic Cooperation and Development OECD, OECD Publishing, Available at: *http://www.planejamento.gov.br/secretarias/upload/Arquivos/seges/arquivos/OCDE2011/OECD_Rethinking_ Approaches.pdf*
35. Palanisamy, R. and Mukerji, B. (2012) 'Security and Privacy issues in e-Government'. *IGI Global*: 236–248.
36. Parent, M., Vandebeek, C. A. and Gemino, A. C. (2004) 'Building Citizen Trust through e-Government'. *Proceedings of the 37th Hawaii International Conference on System Sciences*: 1–9. Available at: *http://csdl.computer.org/dl/proceedings/hicss/2004/2056/05/205650119a.pdf*
37. Reddick, C. G. (2010a) 'Politics, Democracy and e-Government: Participation and Service Delivery'. Hershey, PA: Information Science Reference.
38. Reddick, C. G. (2010b) 'Citizens and e-Government: Evaluating Policy and Management'. Hershey, PA: Information Science Reference.
39. Roebuck, K. (2011) 'Federated ID Management'. Tebbo Publishing.
40. Serageldin, I. (2011) 'Science and the Arab spring. Issues in Science and Technology'. Available at: *http://www.issues.org/27.4/p_ serageldin.html*
41. Sheldon, F. T. and Vishik, C. (2011) 'Moving toward trustworthy systems: R&D Essentials'.*Computer* 44 (9): 31–40. Available at: *http://www.computer.org*
42. Suomi, R. and Tähkäpää, J. (2002) 'The Strategic Role of ICT in the Competition Between Public and Private Health Care Sectors in the Nordic Welfare Societies – Case Finland'. *Proceedings of the 35th Hawaii International Conference on System Sciences*. Available at: *http://www.computer.org/comp/proceedings/hicss/2002/1435/06/14350145b.pdf*
43. Tassabehji, R. and Elliman, T. (2006) 'Generating Citizen Trust in e-Government using a Trust Verification Agent: A research note'. *European and Mediterranean Conference on Information Systems (EMCIS)*. July 6–7, Alicante, Spain.
44. Torres, L., Vicente, P. V. and Royo, S. (2005) 'E-government and the Transformation of Public Administrations in EU Countries: Beyond NPM or just a Second Wave of Reforms'? Available at: *http://www.dteconz.unizar.es/DT2005-01.pdf*
45. UNDP (2010) 'United Nations Global e-Government Survey 2010'. Available at: *http://www2.unpan.org/egovkb/documents/2010/E_Gov_2010_ Complete.pdf*
46. UNDP (2012) 'United Nations e-Government Survey 2012: e-Government for the People'. Available at: *http://unpan1.un.org/intradoc/groups/public/documents/un/unpan048065.pdf*
47. Varian, H. R., Farrell, J. and Shapiro, C. (2005) 'The Economics of Information Technology'. New York: Cambridge University Press.

48. Warkentin, M., Gefen D., Pavlou P. A. and Rose, G. (2002) 'Encouraging Citizen Adoption of e-Government by Building Trust'. *Electronic Markets* 12 (3): 157–162.

49. Windley, P. J. (2005) 'Digital identity' (1st Edition). Sebastopol, CA: O'Reilly.

50. Yee, G., El-Khatib, K., Korba, L., Patrick, K. A. S., Song, R. and Xu, Y. (2005) 'Privacy and Trust in e-Government'. *Electronic Government Strategies and Implementation*: 145–190.

e-Government in Arab countries: a six-stage road map to develop the public sector

Abstract: Governments the world over are competing with each other to be in a leading position in the arena of e-government. e-Government is seen as a path to modernisation and rendering more efficient and effective public sector services. Recent practices in the field have focused on bringing the government closer to the people. As such, governments worldwide have adopted various government-to-citizen (G2C) e-government models in an attempt to improve and provide round the clock availability of all government public services. This article provides a review of the current e-government field with a focus on Arab countries. We present a conceptual six-stage road map that illustrates our account of how Arab countries should prioritise their e-government short and mid-term efforts. It is a simplified model that represents mega-functions that governments need to bear in mind when addressing the changing development needs of the globalised world we live in today. We argue that the stages of the proposed road map have the potential to support the development of the public sector and the emergence of the Arab bloc as strong, revolutionised, citizen-centric governments.

Keywords: e-Government, citizen centricity, participation, inclusion, Arab countries.

1. Introduction

'e-Government has been a bipartisan effort, today it becomes bicameral as well... The era of big government is over we are committed to a smaller, smarter government. New information technologies are tools to help us achieve these goals'.

Joseph Lieberman

It comes as a paradox when we attempt to examine the definitions of 'government' and 'e-government'. The commonly recognised definition of government refers to it as being 'the act or process of governing; specifically: authoritative direction or control' (Merriam-Webster, 2011). All other definitions expressed in dictionaries and the literature indicate government to be an 'authority', 'enforcer', 'controller' and so on (MWD, 2011).

On the other hand, e-government is associated with terms such as development, services, access and relationships. As per the United Nations' definition of the concept of e-government, or 'digital government', it can be understood as 'the employment of the Internet and the worldwide-web for delivering government information and services to the citizens' (United Nations, 2006; AOEMA, 2005).

Therefore, e-government in its essence is about transforming relations with citizens, businesses, and other arms of government with the objective to enhance the overall efficiency and effectiveness of service delivery in the public sector (Hai, 2007; World Bank, 2012). See also Figure 4.1.

e-Government is more to do with the enablement and facilitation of citizen relationships with the government as it takes the process of governance closer to the people. After all, governments are meant to

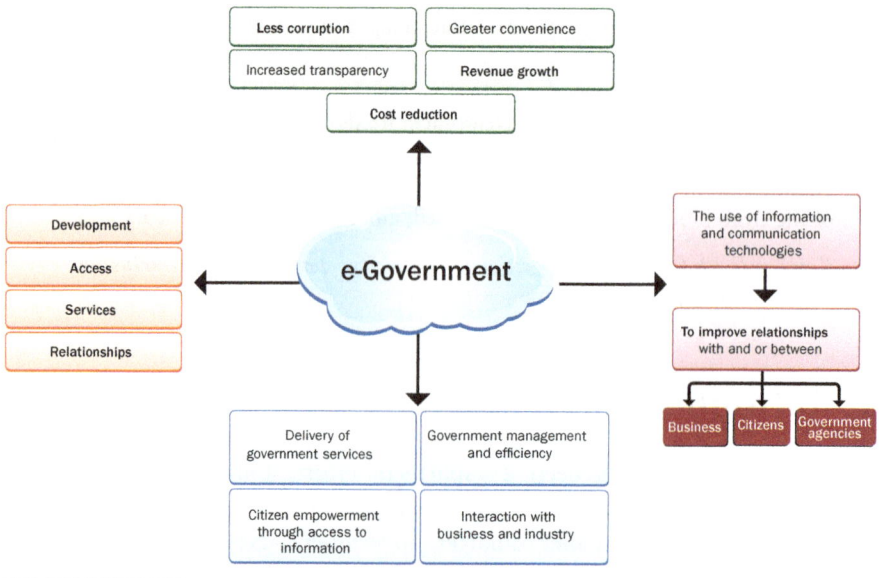

Figure 4.1 The e-government context

improve life quality and instil a sense of inclusion among its population. Accordingly, e-government enables and facilitates this specific objective through the delivery of services to the citizens and residents at their personal convenience, allowing secure personal transactions with the government with a choice of channels and time, thereby bringing them closer to the government authorities. The potential is endless.

The aim of this article is to provide an overview of the existing body of knowledge in the field of e-government, and relate it to the discussed topics around citizen centricity and the delivery of public services. The discussion in the second part of the article focuses more on e-government practices in Arab countries. The major contribution of the article is presented in the form of a six-stage road map that illustrates our account of how Arab countries should prioritise their e-government short and mid-term efforts. The road map can be used as a tool for progress monitoring and measurement. Its stages are envisaged to have the potential to support the development of the public sector and address the changing needs of the globalised world we live in today.

The article is organised as follows: Section 2 outlines the concept of citizen-centric e-government. It emphasises the need to adopt a strategic planning model to ensure successful e-government implementation. Section 3 presents a number of e-government maturity models to illustrate e-government in the context of its progressive phases. Section 4 emphasises the need to develop information societies to ensure 'inclusive growth' and shift towards citizen-centric systems. Section 5 presents some recent statistics reported by the United Nations on the development index and evolution of e-government globally. Section 6 sheds light on e-government maturity in the context of Arab countries. In section 7, the proposed roadmap is presented and its phases are discussed in detail. Finally, some reflections and thoughts are presented in section 8, where the article is concluded.

2. Citizen-centric government

'Government of the people, by the people, for the people, shall not perish from the earth'. Abraham Lincoln (1863)

Citizen centricity is the crux of governance in modern government terms. It redefines the parameters on which governments interact with their citizens. It also calls on governments for more openness, transparency and collaboration. At its most fundamental level, citizen centricity is a mind shift from an 'institution-centred' view of government to a 'citizen-centred' view of government (Suthrum and Phillips, 2003).

Citizen centricity is not an option for governments in today's vocabulary. In the face of political and economic unrest around the globe, a need arises to redefine the balance between the governing parties and their constituents, in a citizen-centred paradigm that will bring the human factor back into the centre of the equation and will focus on the citizens and their well-being (Rahav, 2012). It is about changing priorities and opting to put citizens and their well-being at the heart of government policies and systems (Bhatnagar, 2008; Kearney, 2011; OECD, 2008; Rahav, 2012).

Current research indicates the need for the effort required to expand the view of the future. It prompts governments to be more practical in how they exploit new opportunities for innovation and transformation to shape their strategic priorities and meet their citizens' needs (Di Maio et al., 2005). Numerous models have been developed to support governments in this domain of application. They present conceptual frameworks to transform government infrastructure and enable citizens to become more participative and inclusive in the governance process itself.

However, practices in the field of e-government point out the difficulty of developing systems that can meet today's increasingly changing and complex landscape (Goldkuhl, 2012). The literature surrounding e-government in general demonstrates that governments need to follow a strategic planning model to ensure successful implementation (Homburg, 2008; Lowery, 2001; Otenyo and Lind, 2011; Sherry et al., 2012). A lack of such strategies is argued to be among the primary causes for existing gaps in coordination and/or communication between various stakeholders and initiatives. It is also attributed to the existence of so-called 'islands of automation' and 'stovepipes' within and between levels of government (Seifert and McLoughlin, 2007).

2.1 Gartner 2020 government scenario planning tool

One often cited scenario-planning tool in the literature for creating long-term government strategies is depicted in Figure 4.2. This was developed by the Gartner research group. The model gives a picture of how governments will use and be shaped by technology in 2020. The framework uses two primary driving forces.

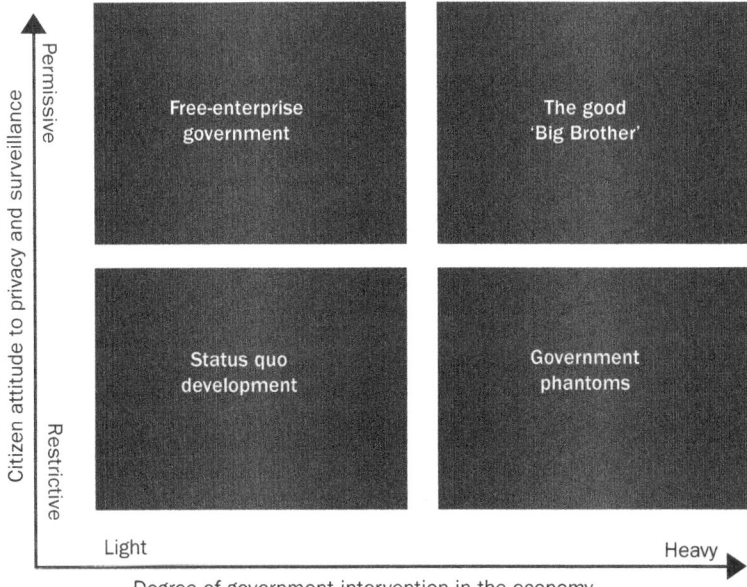

Figure 4.2 Gartner 2020 government scenarios

The first force is related to the degree of government intervention in the economy, and captures the different attitudes that governments can take vis-à-vis the regulation of economy. The second driving force relates to citizen attitude to privacy and surveillance, and ranges from governments with the freedom to access citizens' personal information to situations in which laws protect citizens' private information.

In the scenario of low government intervention and restrictive regimes, the development of the economy is nearly defunct and remains a status quo. In a case when the government intervenes heavily and citizen participation is highly restrictive, the development is purely a function of the government initiatives. On the other hand, low participation from the government, combined with a free and highly permissive environment represents a pure capitalist form of government where the development is determined solely by free enterprise. An all-round development, well-regulated and participative, is the 'Good Big Brother' approach where the government and citizens participate alike in determining the development of the nation.

In practice, we may see all four scenarios to some extent carried out throughout the world, with each country adopting transformation models that better fit their national priorities and policies. In fact, and in the past few years, many countries have launched different initiatives for the transformation of their governments to increase the inclusiveness of their citizens and residents, all striving to reach out to their citizens and improve their quality of life (Al-Khouri, 2012; CARICOM, 2009; Nordfors et al., 2006). The essence of these initiatives is to 'take the government to the citizen', and move forward from the conventional concept of government offices.

In such transformed operating models, the government is no longer confined to four walls, filing cabinets and service counters. It also means that governments can be available on a 24/7 basis, (e.g. over the Internet and through mobile phones, public kiosk machines, digital TV, and call centres, as well as through personal computers). In such operating models, governments attempt to create service-oriented architectures (SOA)* develop a single platform through which public services are provided.

Prior to digging deeper in the literature to explore governments' maturity, we attempt in the next section to shed light and define the term 'maturity' in the context of e-government and existing developed models to gauge its progress.

3. e-Government maturity models

'It is the framework which changes with each new technology and not just the picture within the frame'.

Marshall McLuhan

e-Government literature in general classifies its focus areas into three primary groups: citizens, businesses, and the government. The different models indicate the development of relationships between the government and citizens (G2C), government and businesses (G2B), and between government agencies (G2G). See also Figure 4.3.

---

*In software engineering, a Service-Oriented Architecture (SOA) is a set of principles and methodologies for designing and developing software in the form of inter-operable services. SOA design principles are used during the phases of systems development and integration to define how to integrate disparate applications for a Web-based environment and use multiple implementation platforms. SOA is not just an architecture of services seen from a technology perspective, but the policies, practices, and frameworks by which we ensure the right services are provided and consumed.

		To (destination)		
		G (Government)	B (Business)	C (Citizen)
From (origin)	G	G2G Back office conversion	G2B Front office input	G2C Front office input
	B	B2G Front office input	B2B (E-business)	B2C (E-business)
	C	C2G Front office input	C2B (E-business)	C2C (E-community)

Figure 4.3 Electronic interactions of e-government (Song, 2006)

3.1 The Forrester e-Government Maturity Continuum

One illustrious transformation model that depicts the evolution of e-government was developed by Forrester Research and called the e-Government Maturity Continuum. Today, e-government initiatives fall somewhere within the three phases of the framework, namely access era, interaction era, and integration era. See also Figure 4.4. The framework is used to depict the evolution and transformations that took place over a period of 20 years starting in 1993, the year in which major e-transformations for citizen-centric governments really began.

The era of access: in this first phase of citizen-centric government initiatives, citizens are able to access government information online. This is the information dissemination and information sharing phase. The idea is to provide real-time information to reach out to the confidence of the people.

The era of interaction: this phase is to allow transactions. In this stage of development, there is some level of transactions taking place between the government and the citizen. Many governments today have adopted

---

* Access and Identity Management (AIM) is a term that describes the management of individual identities, their authentication, authorisation, and privileges/permissions within or across system and enterprise boundaries with the goal of increasing security and productivity while decreasing cost, downtime, and repetitive tasks.

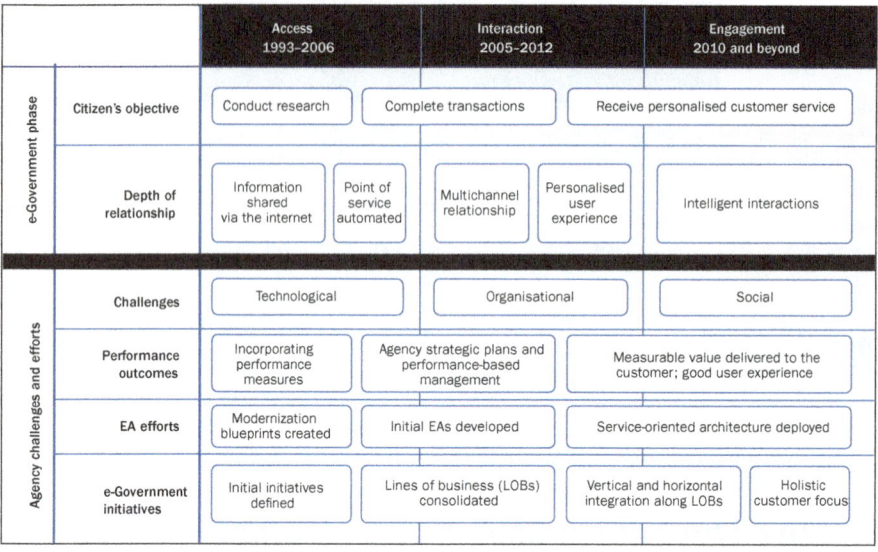

		Access 1993–2006		Interaction 2005–2012		Engagement 2010 and beyond	
e-Government phase	Citizen's objective	Conduct research		Complete transactions		Receive personalised customer service	
	Depth of relationship	Information shared via the internet	Point of service automated	Multichannel relationship	Personalised user experience	Intelligent interactions	
Agency challenges and efforts	Challenges	Technological		Organisational		Social	
	Performance outcomes	Incorporating performance measures		Agency strategic plans and performance-based management		Measurable value delivered to the customer; good user experience	
	EA efforts	Modernization blueprints created		Initial EAs developed		Service-oriented architecture deployed	
	e-Government initiatives	Initial initiatives defined		Lines of business (LOBs) consolidated		Vertical and horizontal integration along LOBs	Holistic customer focus

Figure 4.4   Citizen-centric government phases

Source: Forrester Research, Inc

Access and Identity Management (IAM)* systems to enable and facilitate such transactions.

The era of engagement: this is the last phase of citizen-centric e-government maturity. It is about enhancing the participation of the citizens in the different aspects of government decision-making in a proactive rather than an impersonal way. Decision-making is no longer the realm of a handful of bureaucrats or a few politicians. Decision-making is heavily influenced by the citizens who are able to use their voices to shape the policies that govern their lives. At this stage of development, government websites are topically or user-group oriented, enlightened by insights about the constituents they serve. These sites span multiple agencies and multiple levels of government, are more intuitive to use, and reach across multiple channels seamlessly.

Two other famous e-government models are depicted in Figures 4.5 and 4.6. Both of the models complement Forrester's e-Government Maturity Model.

3.2 Layne and Lee's e-Government Model

Layne and Lee (2001) regarded e-government as an evolutionary phenomenon and proposed a four-stage model (see Figure 4.5):

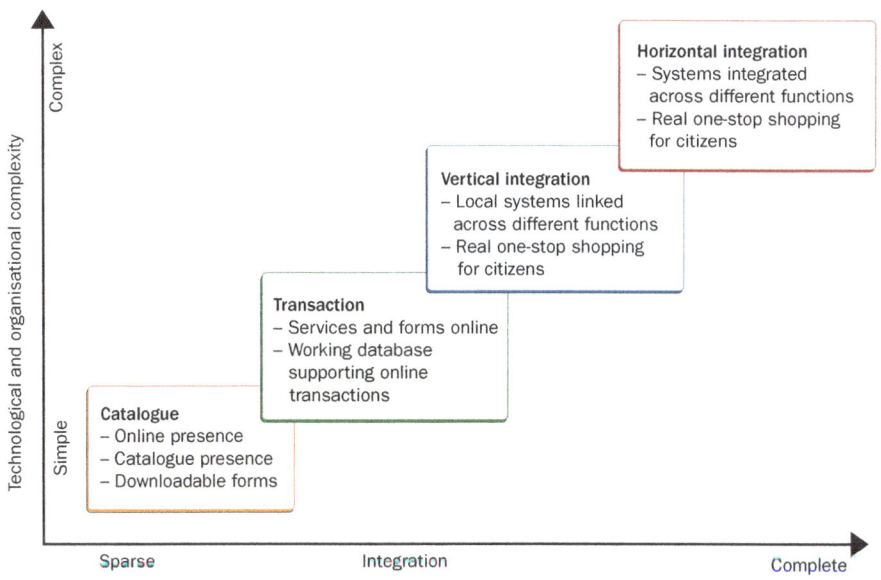

Figure 4.5   Layne and Lee's e-Government Model

1. The catalogue stage delivers some static or basic information through websites.

2. The transaction stage extends the capability of catalogue and enables citizens to do some simple online transactions such as filling government forms.

3. The vertical integration stage initiates the transformation of government services rather than automating its existing processes. It focuses on integrating government functions at different levels, such as those of local governments and state governments.

4. The horizontal integration stage focuses on integrating different functions from separate systems to provide users with a unified and seamless service.

3.3 The Public Sector Process Rebuilding (PPR) Model

Andersen and Henriksen's (2006) Public Sector Process Rebuilding (PPR) model is argued to be an extension of the Layne and Lee Model. (See Figure 4.6). They make a case that the PPR model expands the

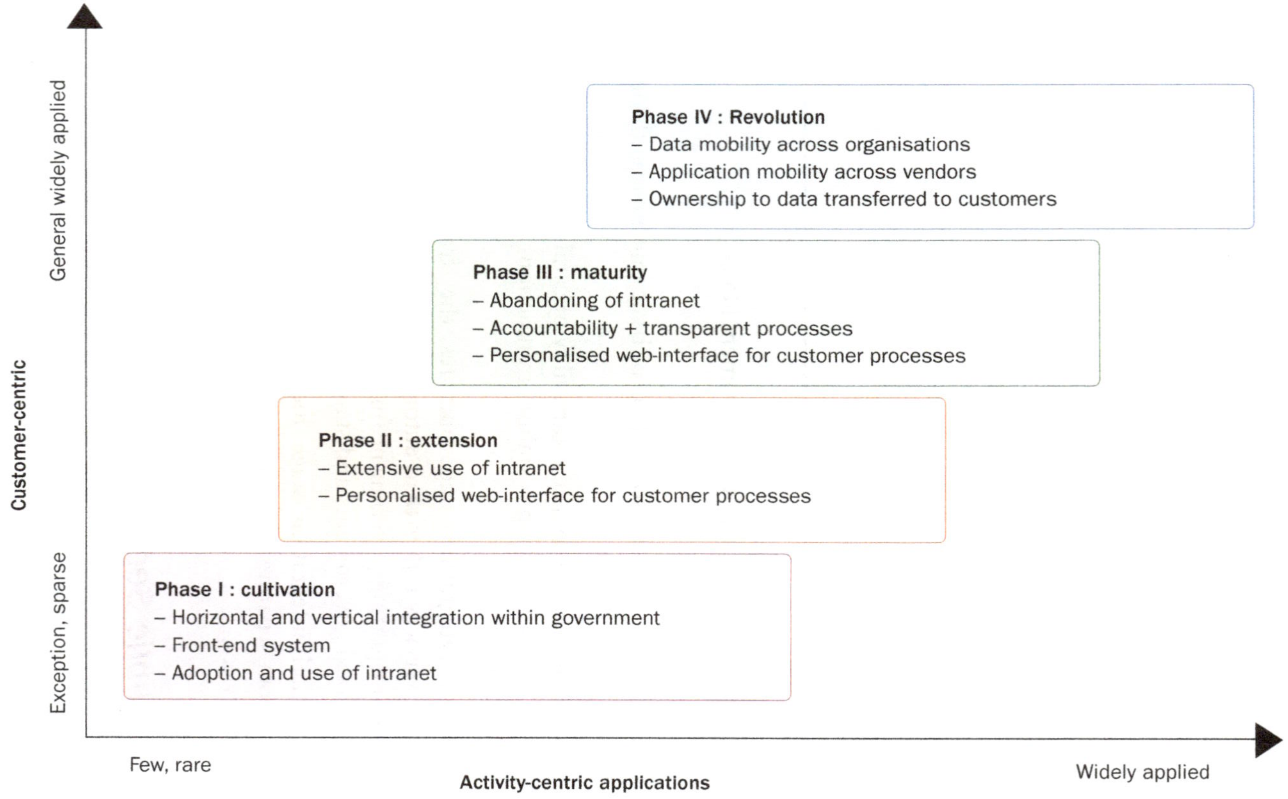

Phase IV : Revolution
– Data mobility across organisations
– Application mobility across vendors
– Ownership to data transferred to customers

Phase III : maturity
– Abandoning of intranet
– Accountability + transparent processes
– Personalised web-interface for customer processes

Phase II : extension
– Extensive use of intranet
– Personalised web-interface for customer processes

Phase I : cultivation
– Horizontal and vertical integration within government
– Front-end system
– Adoption and use of intranet

Widely applied

Activity-centric applications

Few, rare

General widely applied

Exception, sparse

Customer-centric

Figure 4.6   **The Customer and Activity-centric Maturity Model**

e-government focus to include the front-end of government with a focus on the activity and customer-centric approach rather than technological capability.

The PPR model consists of four phases:

1. The cultivation phase shelters horizontal and vertical integration within government, limited use of front-end systems for customer services, and adoption and use of intranet within government. Organisations in this group are not likely to have digital services and will rarely have instant processing capabilities online. Less attention is given to the use of the Internet to increase user frequency, services provided, and/or the quality and speed of services. The downside is that the public institution in this phase will be experienced as inaccessible, have long case processing times, and no accessibility for accessing the processing of requests.

2. The extension phase involves extensive use of the intranet and adoption of personalised web user interfaces for customer processes. It may be characterised by costly user interfaces, no integration with other systems, expensive maintenance, and fading out of old software and data formats. Thus, there are still many manual routines, while the user might find many forms of information where the agency re-directs users to information at other agencies.

3. The maturity phase: organisations in this phase mature and abandon the use of the intranet, have transparent processes, and offer personalised self-service web interfaces for processing customer requests. In this phase, Internet and intranet applications are merged to lower marginal costs for processing the customer requests for services.

4. The revolutionary phase is characterised by data mobility across organisations, application mobility across vendors, and ownership to data transferred to customers.

3.4 Other models for e-government

Several different researchers have developed models to explain the growth of e-government. A short outline about some of these models is provided in Table 4.1.

Despite the various models that have been developed to support e-government progress and maturity, governments in practice have gained limited success in the development of a '24-hour authority'. This is argued to be due to the fact that such initiatives require governments

Table 4.1 e-Government maturity models

Source/proponent(s)	Phases	Description
Chen (2002)	Phase 1: Information Phase 2: Communication Phase 3: Transaction Phase 4: Transformation	1. Government 'information' is created, categorised, and indexed and delivered to its citizens through the Internet. 2. e-Government services support two-way 'communication,' with citizens communicating requests through web forms, email, or other Internet media. 3. 'Transaction' services between citizens and governments are supported. Government branches also use the Internet for transactions among themselves. 4. An opportunity for the 'transformation' of government practices and services is exploited. Applications such as e-voting and e-politics that may alter the democratic and political processes are instituted.
Chandler and Emanuel (2002)	Stage 1: Information Stage 2: Interaction Stage 3: Transaction Stage 4: Integration	1. Government services are delivered online. One-way communication between government and citizens is put in place. 2. Simple interaction between citizens and governments are supported. 3. Services enabling transactions between citizens and government are supported. 4. Integration of services across the agencies and departments of government are put in place.
Howard (2001)	Phase 1: Publishing Phase 2: Interaction Phase 3: Transaction	1. Information about government activities is available online. 2. Enables citizens to have simple interactions through emails with their governments. 3. Provides citizens with full transaction benefits over the internet with services such as purchasing licenses and permits.

Table 4.1   e-Government maturity models (*Cont'd*)

Source/ proponent(s)	Phases	Description
West (2004)	Phase 1: Billboard Phase 2: Partial service delivery Phase 3: Fully integrated service delivery Phase 4: Interactive democracy with public outreach and accountability.	1. Government websites (usually static at this stage) are used for information display. 2. Government websites have more capabilities and functionalities to include sorting and searching of information. 3. One-stop centre is created with fully integrated online services. 4. Government website develops into a system-wide political transformation, with executable and integrated on-line services. Customised information service is available.
Gartner's group model in Baum and Maio (2000)	Phase 1: Web presence Phase 2: Interaction Phase 3: Transaction Phase 4: Transformation	1. Government uses the web to provide basic information. 2. Government provides a website equipped with search engines, documents downloading capability and emails. 3. Citizens can carry out enhanced online transactions. 4. All government services and processes are integrated, unified and personalised.
Deloitte & Touche (2001)	Phase 1: Information publishing Phase 2: Official two-way transactions Phase 3: Multi-purpose portals Phase 4: Portal personalisation Phase 5: Clustering of common services Phase 6: Full integration and enterprise transformation	1. Government creates websites (static) to provide information to its citizens. 2. Enables customers to have electronic interaction with government services such as television licenses renewal. 3. Enables customers to obtain government services and information from a single point. 4. Government provides customers and its agencies with opportunities to customise portals according to their needs. 5. All government services and processes are clustered so as to provide unified and seamless services to citizens. 6. Government changes its structure to enable the provision of more sophisticated, integrated and personalised services to its citizens.

| Table 4.1 | e-Government maturity models (*Cont'd*) |

Source/ proponent(s)	Phases	Description
UN Public Administration Programme (2010)	Phase 1: Emerging Phase 2: Enhanced Phase 3: Transactional Phase 4: Connected	1. Government provides information and basic services on its web site. 2. Government websites deliver enhanced one-way or simple two-way communication between government and citizens through the use of downloadable forms. 3. Government websites use advanced two-way communication between government and its citizens. The websites process transactions such as e-voting, filling of taxes, and licenses and certificate applications. 4. Government websites change the way it communicates with citizens; they are proactive in requesting opinions and information from their citizens; they create and 'empower' citizens with more voice in decision-making.

to overcome politics and standardise internal processes and data in order to integrate back-office functions across the public sector (OECD, 2009). Some researchers argue that governments need to redefine the term 'access to government' and instead be referred to as 'participation'. It is also argued that governments need to go beyond superficial initiatives, towards a more radical re-engineering of the government processes across its agencies to enable citizen-centric systems and the development of information societies.

4. Information society

'The empires of the future are the empires of the mind'.

'Winston Churchill

Information society is a term associated with the development of a more open, inclusive and sustainable information-based society (see Figure 4.7). It refers to a society where the creation, distribution, diffusion, use,

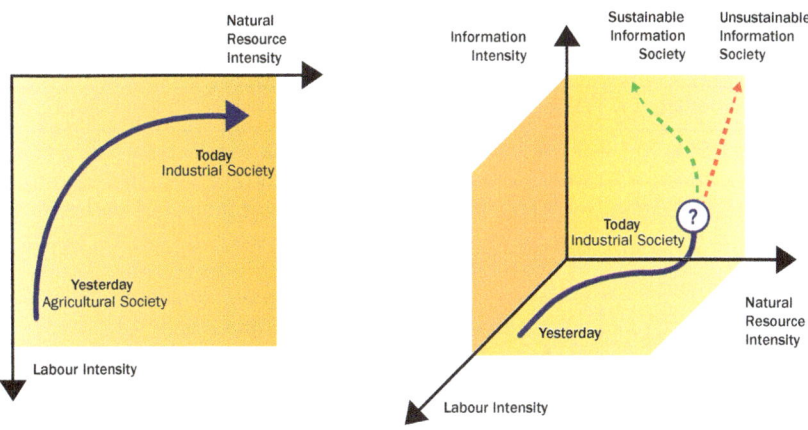

Figure 4.7   Visions of the information society (Hilty, 2011)

integration and manipulation of information is a significant economic, political, and cultural activity (Beniger, 1986). The aim of the information society is to gain a competitive advantage through using information and communication technology (ICT) in a creative and productive way (Feather, 2008; Mattelart, 2003). The knowledge economy is its economic counterpart, whereby wealth is created through the economic exploitation of understanding the different influencing factors and the role of people. People who have the means to partake in this form of society are sometimes called digital citizens. Table 4.2 provides an elaboration of the three main characteristics that symbolise information societies.

Many countries have followed different transformation models to enable their participation in, and development of citizen-oriented systems (Birch, 2002; Gibson et al., 2005; Hayden et al., 2002; Lowndes et al., 2001; Suh, 2007). The following two subsections provide a short overview of relevant government practices to promote citizen participation in, and development of information society, namely the Swedish and European experiences.

4.1 The Swedish 'information society for all' policy

In 2000, Sweden set the policy goal to become the first country to be an 'Information society for all' (European Union, 2012). Since then, the Swedish Government's priority activities have been to enhance public

Table 4.2   Characteristics of the information society

Characteristics	Description
Information is used as an economic resource	Organisations make greater use of information to increase their efficiency, to stimulate innovation and to increase their effectiveness and competitive position, often through improvements in the quality of the goods and services that they produce. There is also a trend towards the development of more information-intensive organisations that add greater amounts of value and thus benefit a country's overall economy.
More intensive use of Information	It is possible to identify greater use of information among the general public. People use information more intensively in their activities as consumers: to inform their choices between different products, to explore their entitlements to public services, and to take greater control over their own lives. They also use information as citizens to exercise their civil rights and responsibilities. In addition, information systems are being developed that will greatly extend public access to educational and cultural provision.
Development of an information sector within the economy	The function of the information sector is to satisfy the general demand for information facilities and services. A significant part of the sector is concerned with the technological infrastructure, the networks of telecommunications and computers. Increasingly, however, the necessity is also being recognised to develop the industry generating the information that flows around the networks: the information-content providers. In nearly all information societies, this sector is growing much faster than the overall economy. The International Telecommunications Union (ITU) estimates that, in 1994, the global information sector grew by over five per cent while the overall world economy grew by less than three per cent.

confidence in IT, and help to improve user skills and foster access to IT services. According to the Sweden 24-hour Public Administration Strategy, public information and services should, to the maximum degree, be electronically available 24 hours a day, seven days a week. Another major aim of the strategy was to strengthen democracy by enhancing transparency and citizen participation in the policy and decision-making processes. The strategy for delivery was based on the Swedish decentralised model for public administration. Next to small policy ministries, a large number of agencies are responsible for implementing government policy. The agencies are managed by a system of performance management, where the government sets targets, allocates resources and appoints managers while following up and evaluating results.

Sweden remains the most competitive economy as measured by the European Union's (EU) own competition benchmark, the Lisbon criteria,* followed by Finland, Denmark and the Netherlands, according to the World Economic Forum's Lisbon Review 2010 (World Economic Forum, 2010). Figure 4.8 depicts a chart that compares Sweden's performance, vis-à-vis the US and East Asia benchmarks. In the figure, Sweden's performance is represented by a blue line, that of the US is in grey, and that of East Asia is in black. Dimensions in which the blue line extends further out than that of the US or East Asia indicate areas where Sweden outperforms these comparators.

4.2 The European information society development framework

Europe's Information Society policies are brought together under the i2010 Initiative, the EU framework for addressing the main challenges and developments in the information society and media sectors in the years up to 2010 (European Commission, 2010). The initiative promotes an open and competitive digital economy, research into information and communication technologies, as well as their application to improve social inclusion, public services and quality of life. One of the frameworks that has been developed to trigger debate and discussion in the European Union is depicted in Figure 4.9.

---

*The World Economic Forum's study is a biennial review series that assesses the progress made by EU Member countries in the far-reaching goals of the EU's Lisbon Strategy of economic and structural reforms. In addition to assessing the performance of 27 existing EU Members, it also measures the competitive performance of EU candidates and potential candidate countries.

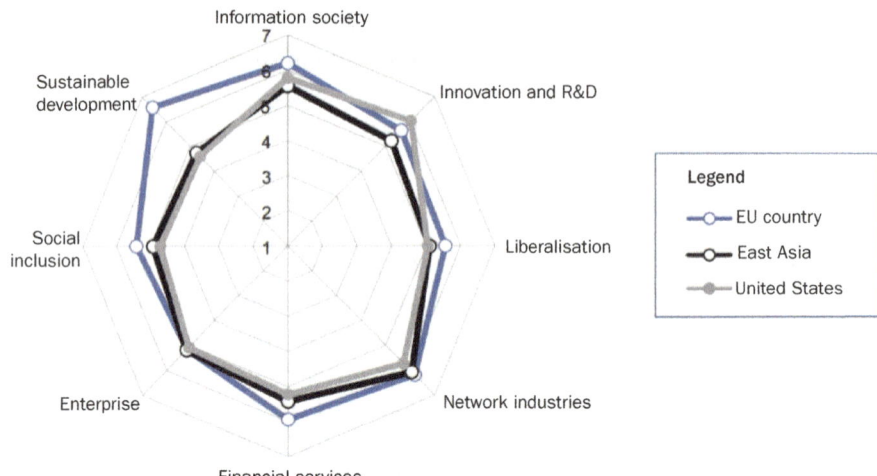

Information society

Legend
- EU country
- East Asia
- United States

Sustainable development

Social inclusion

Enterprise

Financial services

Innovation and R&D

Liberalisation

Network industries

Figure 4.8 Swedish performance benchmark vis-à-vis the US and East Asia

	Vertical sectors: main axis addressed to government, people, enterprises and urban environment			
Horizontal layers	**Government**	**People**	**Enterprises**	**Urban Environment**
Content	e-governance	Community development	e-inclusion and networking for SMEs	Local environment and sustainable development
Applications	e-government	e-inclusion for specific groups	Entrepreneurship and the local economy	Culture and tourism
Generic services	Public services online	Education and training	Development and support centres	Urban management (i.e. GIS, transport, land management)
Network	e-participation	Entertainment and culture	e-business and e-commerce	
		e-Health	Lifelong learning	
	Access	**Knowledge**	**Skills**	**Sustainable management**

Diagonal areas:
Awareness raising, regulatory framework, public-private partnership, financial engineering, social capital, resource management

Main dimensions and axis for collaboration at the city level

Figure 4.9 A Framework for co-operation in developing the information society in Europe (digital inclusion and capacity building): an instrument to stimulate debate

The origin of the framework goes back to 1995, when the European Union launched an initiative for decentralised cooperation networks between cities and towns cutting across geographical boundaries. The framework seeks to provide complete inclusion of citizens in different stakeholder roles for promoting best practices for the cooperation between towns and cities in building information societies. It depicts a multi-dimensional matrix of the stakeholders in the cooperation, namely government, people, enterprises, and at the decentralised level the urban environment.

Government focus areas for citizen inclusion and cooperation would be e-government, e-services, e-participation, and online public services. These result in higher transparency and greater efficiency in service delivery. This efficiency, productivity and citizen satisfaction are determined by the accessibility of the services. The access mechanism itself is determined by the content of the services, the type of applications available for the services and their delivery, and the type of networks enabling this access. Participation is enhanced by the knowledge and skills available using different access channels. All the systems need sustainable management to ensure continual and continuous improvement in the practices to be adopted and followed.

People factors include special interest groups, education and training, health, entertainment and overall community development. Practices attributed to these factors lead to an improvement in quality of life. Enterprises contribute heavily to the economic development.

SMEs, entrepreneurship, development and support centres for economic growth, and facilitation of e-commerce are the focus areas of enterprises. The importance of access mechanisms, skills sets, knowledge and sustainable management should be underlined in the development of best practices.

Last but not the least is the urban environment in the cooperation and citizen inclusion framework. The urban environment constitutes the key decentralised governance domain contributing to citizen inclusion, directly leading to higher cooperation between diverse geographic locations.

Having said this, the next section presents some recent statistics reported by the United Nations on the development index and evolution of e-government worldwide, and we shall take a closer look at e-government in Arab countries in the subsequent section.

5. The United Nations e-government assessment

'There is no discipline in the world so severe as the discipline of experience subjected to the tests of intelligent development and direction'.

John Dewey (1859–1952)

The United Nations has been tracking the development and evolution of e-governments for the past decade. The UN e-government survey provides a bi-annual assessment of national online services, telecommunication infrastructure and human capital of 192 member states. The UN's e-government maturity model indicates that countries go through primarily four phases of maturity, as depicted in Figure 4.10.

As per the latest e-government survey in 2012, countries that have advanced to higher growth levels on their e-government projects are the ones with relatively high Web measurement and online service index scores. South Korea was rated first for its comprehensive online infrastructure and user involvement level, with a development index of 0.9283, followed closely by the Netherlands with 0.9125. The rest of the top ten countries included the UK, the US, Denmark, Norway, France, Sweden, Finland, and Singapore. It is interesting that these countries have very little differences in the composite index that determines their ranking. All top ten countries have similarities in how they have implemented e-government. What is more interesting is the list of countries that are identified as the emerging leaders in e-government development (see Figure 4.11). This is the list where each country is distinct in its demographics and political content.

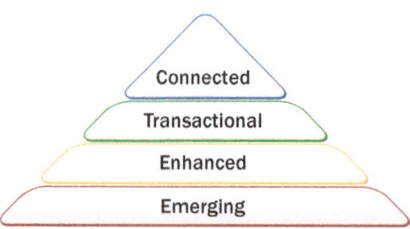

Figure 4.10   The UN's four stages of online service growth

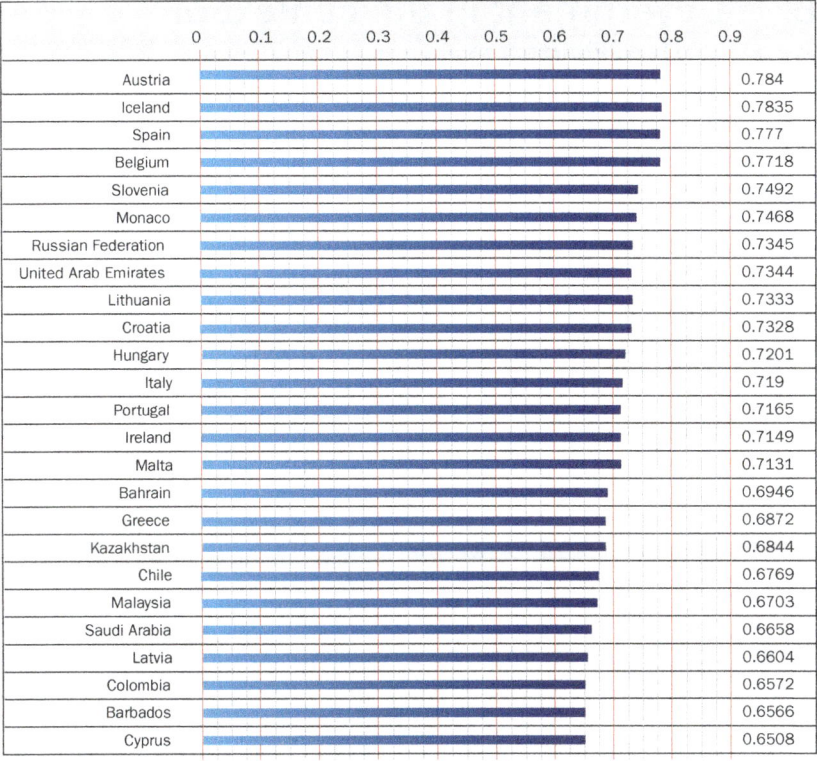

Figure 4.11 Emerging leaders of e-government development index (United Nations, 2012)

In addition, the list of emerging leaders in e-government development brings another very interesting observation to the fore. The list prominently features three Arab countries, namely the UAE, Bahrain and Saudi Arabia. These countries are leading the transformation of governments in the Middle East region. With increasing use of Internet, social networking, and communications technology, combined with a fast-growing educated population, these Arab countries are attempting to create for themselves examples of citizen-centric governments. The next section further elaborates on this.

6. e-Government in the context of Arab countries

'No one ever teaches well who wants to teach, or governs well who wants to govern'.

Plato

According to a United Nations 2012 survey, the average e-government development index in Western Asia, which largely comprises Arab nations, is well above the world average; 0.5547 to 0.4882 in 2012 (UNPAN, 2012) as can be seen in Figure 4.12. This shows an increase of six per cent over the ratio reported in 2010, as they have shown more growth according to e-government development indexes.

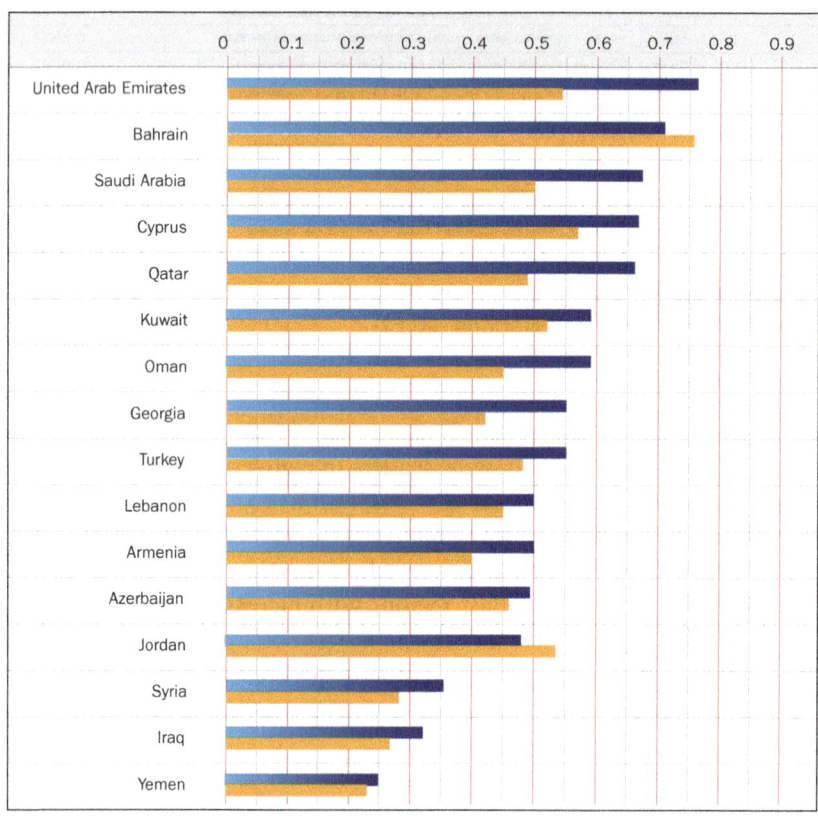

Figure 4.12 Arab nations' e-government development index (United Nations, 2012)

This growth has been enabled by double-digit growth rates in the past few years in the telecommunications industry in Middle-Eastern countries. Telecom companies like ETISALAT (UAE), QTEL (Qatar), BATELCO (Bahrain), STC, Mobily, and Zain (Saudi Arabia), have transformed the telecommunications landscape in the region. They moved from basic telephony providers to providing converged telecommunication services with 3G and 4G services on mobile phones, and to providing fibre-optic networks across their countries, resulting in faster roll-out and more reliable Internet connections.

Figure 4.13 provides a quick glance at the latest statistics published by Internet World Statistics, which reveal that Internet penetration in the Middle-East is higher than the rest of the world (IWS, 2012). For instance, higher Internet penetration has enabled a larger number of countries in the region to host government web portals as eminent channels to disseminate information. Furthermore, mobile telecommunications are enabling these governments to deliver services through diverse channels. According to the United Nations (2011), the United Arab Emirates, Kuwait, and Estonia have made the same amount of progress in less than two years.

6.1 Arab countries and government portals

All Arab bloc countries have strong leadership backing and funded initiatives to modernise existing public service delivery infrastructures. As some Arab countries continue their investment in digital knowledge-based economies, the regional ICT infrastructure is growing exponentially. Internet penetration and household usage of technology for communication shows a year after year growth in double-digit figures.

Internet Users in the Middle East and in the world						
Middle East region	Population (2011 Est.)	Pop.% of World	Internet Users 31 Dec 2011	Population % (Penetration)	Users % World	Facebook 31 Dec 2011
Total Middle East	216,258,843	3.1%	77.020.995	35.6%	3.4%	18,241,080
Rest of the world	6,713,796,311	96.9%	2,190,212,747	32.6%	96.6%	780,851,080
World total	6,930,055,154	100%	2,267,233,742	32.7%	100%	799,092,160

Notes:
(1) Internet usage and population statistics for the middle ease were updated as of December 31, 2011
(2) Population numbers are based on data contained in the US Census Bureau
(3) The most recent Internet Stats come mainly from data published by Nielsen online, ITU, Facebook, and other trustworthy sources,
(4) Data on this site may be cited, giving due credit and establishing an active link back to internetWorldStats.com

Figure 4.13 The status of communications and service delivery channels in Arab countries

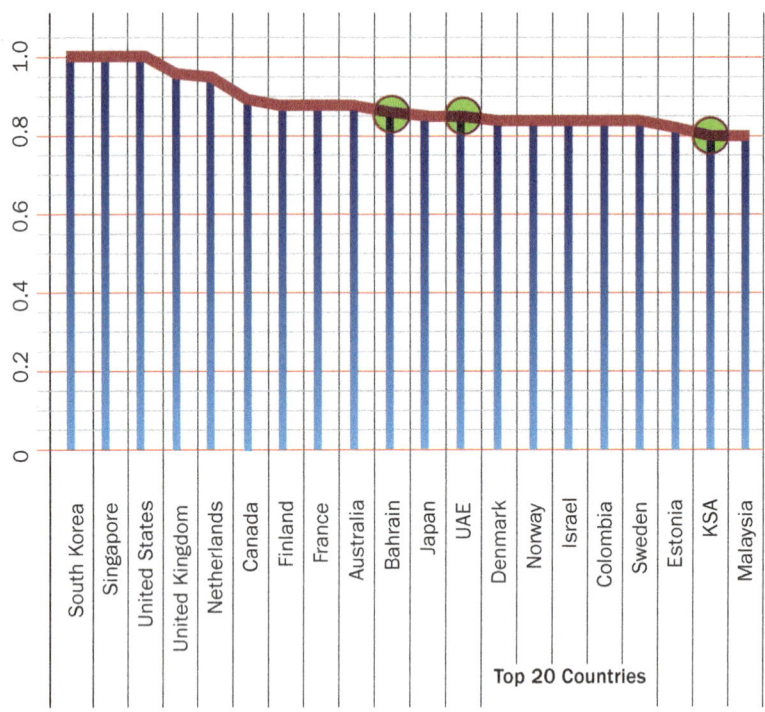

Figure 4.14 Top 20 countries in online service delivery (United Nations, 2012)

The UAE, Bahrain and Saudi Arabia are considered to be leading the Arab countries in the delivery of online public services. The UAE and Bahrain in particular are setting examples in taking a whole-government and people-centred development approach. Some of the key initiatives launched in recent years to support citizen-centric e-government development in these countries are related to network readiness, infrastructure readiness, service availability, citizen inclusion, and the development of a national identity management infrastructure.

The UNPAN e-Government Survey 2012 stated that the, 'United Arab Emirates (0.7344) is especially notable as it advanced 21 positions to the

Country/Economy	2010–2011			2009–2010	
	Rank		Score	Rank	Score
Sweden	1	▪	5.6	1	5.65
Singapore	2	▪	5.59	2	5.64
Finland	3	▲	5.43	6	5.44
Switzerland	4	▪	5.33	4	5.48
USA	5	▪	5.33	5	5.46
Taiwan, China	6	▲	5.3	11	5.2
Denmark	7	▼	5.29	3	5.54
Canada	8	▼	5.21	7	5.36
Norway	9	▲	5.21	10	5.22
South Korea	10	▲	5.19	15	5.14
Netherlands	11	▼	5.19	9	5.32
Hong Kong	12	▼	5.19	8	5.33
Germany	13	▲	5.14	14	5.16
Luxembourg	14	▲	5.14	17	5.02
UK	15	▼	5.12	13	5.17
Iceland	16	▼	5.07	12	5.2
Australia	17	▼	5.06	16	5.06
New Zealand	18	▲	5.03	19	4.94
Japan	19	▲	4.95	21	4.89
France	20	▲	4.92	18	4.99
Austria	21	▼	4.9	20	4.94
Israel	22	▲	4.81	28	4.58
Belgium	23	▼	4.8	22	4.86
UAE	24	▼	4.8	23	4.85
Qatar	25	▲	4.79	30	4.53
Estonia	26	▼	4.76	25	4.81
Malta	27	▼	4.76	26	4.75
Malaysia	28	▼	4.74	27	4.65
Ireland	29	▼	4.71	24	4.82
Bahrain	30	▼	4.64	29	4.58
Cyprus	31	▲	4.5	32	4.48
Portugal	32	▲	4.5	33	4.41
KSA	33	▲	4.44	38	4.3

Figure 4.15 World Economic Forum Global Information Technology Report 2011

ranking of 28th globally and fifth in Asia. The rapid progress of the UAE is a best practice case highlighting how effective e-government can help support development. With double the population and three-quarters of the GDP per capita, the UAE has achieved around the same level of online services as those offered in Norway, a global leader in eighth position'.

This is validated by the Global Information Technology Report 2011 of the World Economic Forum that provides an index for the overall readiness of the world nations for e-government maturity (Dutta and

Bilbao-Osorio, 2012). The following Figure from this report shows the achievements of the Middle Eastern nations in their endeavours and quest for e-government.

When we compare this table with the readiness scores sub-indexes, where UAE ranks fifth on individual readiness and third on government readiness, it shows a clear intention of making itself an example of a successful e-government case study. This intention can be generalised to the region itself. Qatar, Bahrain, Oman, Saudi Arabia, Jordan, Lebanon, Yemen, Egypt, and Tunisia rank among the top 100 countries in e-government development. All development indexes in the Global Information Technology Report 2011 indicate that the regional average is well above the world average (see figure 4.16).

Taking a cue from these leaders, the region has shown remarkable advancement in the adoption of citizen-centric e-government programs. According to the report released by the Arab Advisors Group (Al Borgan, 2011), 15 Arab countries host 21 government portals. This report analysed 19 countries listed alphabetically as Algeria, Bahrain, Egypt, Iraq, Jordan, Kuwait, Lebanon, Libya, Mauritania, Morocco, Oman, Qatar, Palestine, Saudi Arabia, Sudan, Syria, Tunisia, UAE and Yemen.

The most evolved government portals, such as those in the UAE, Bahrain, Qatar, and Saudi Arabia, enable many citizen transactions online. All these portals provide comprehensive information and are following various transformation models to provide transactional services, while the leaders are moving in the direction of whole government. According to the report, the Arab Advisors Group report 20 per cent of these portals provide mobile messaging services, while 65 per cent of the portals are already delivering transactional services. With the widespread usage of mobile phones, there is a clear intention of Arab countries' governments to increase the number of services through more mobile channels.

The available technologies today enable citizen services provided through multiple delivery channels: Internet, mobile phones, help desks/ contact centres, hybrid/composite delivery channels, while conventional service counters are the currently used channels. Each channel serves different contact purposes, though not all of them are equally effective. This takes us to a presentation of our proposed model in the next section.

Economy Country	Overall NRI	Environment			Readiness			Usage			Number of times in	
		Market	Political and Regulatory	Infrastructure	Individual	Business	Government	Individual	Business	Government	Top ten	Top three
Sweden	1	7	2	2	-	2	8	1	6	-	7	4
Singapore	2	5	1	-	1	5	1	7	10	3	8	4
Finland	3	6	4	9	3	3	10	2	8	-	8	3
Switzerland	4	2	6	3	-	1	-	9	5	-	6	3
USA	5	-	0	5	-	6	-	-	3	4	4	1
Taiwan, China	6	-	-	-	-	-	5	-	1	2	3	2
Denmark	7	-	-	10	9	9	-	5	-	9	5	-
Canada	8	4	-	4	6	-	-	-	-	5	4	-
Norway	9	8	8	8	-	-	-	10	-	-	4	-
Korea, Rep.	10	-	-	-	-	-	-	4	2	1	3	2
Hong Kong SAR	12	1	-	-	2	-	-	-	-	7	3	2
Luxembourg	14	3	5	-	-	-	7	3	-	-	4	2
Iceland	16	-	-	1	4	-	-	6	-	-	3	1
New Zealand	18	-	3	-	-	-	-	-	-	-	1	1
UAE	24	-	-	-	5	-	3	-	-	-	2	1
Qatar	25	10	-	-	10	-	2	-	-	-	3	1

Figure 4.16 Development indexes of countries in the Global Information Technology Report 2011

Source: (Dutta and Bilbao-Osorio, 2012)

7. A potential road map of e-government in Arab countries

'To effectively communicate, we must realise that we are all different in the way we perceive the world and use this understanding as a guide to our communication with others'.

Anthony Robbins

Based on our review of the literature and the information we gathered from our experience in the field of e-government, we developed a six-stage road map that provides guidance on how Arab countries should plan and develop their e-government initiatives. This road map provides a more focused view of what the e-government journey needs to concentrate on.

The proposed road map consists of six main phases. The first phase is more about the transformation and automation of back offices and administrative processes to deliver basic services efficiently to their citizens. The second phase is concerned with the unification of service delivery architectures to enable the delivery of government services through multiple channels, i.e. Internet, kiosk machines and so on. The third phase is concerned with providing digital identities to populations to facilitate the development of innovative e-services. This is followed by the integration of service channels in the fourth phase to provide a highly satisfactory user experience. The fifth and sixth phases of the model entail inter-operability across the vertical and horizontal dimensions of the government, leading to regional and international connectivity.

The model is simplified to support key decision-makers to focus on outcomes rather than the current activity-based and output approach that is currently followed in practice. Activities and output relate to 'what we do'. Outcomes refer to 'what difference we make'. On the proposed road map the completion of the phase should indicate that the expected outcomes have occurred. It can be used as a tool for progress monitoring and measurement. We argue that the stages of the proposed road map have the potential to support the development of the public sector. The next sub-sections discuss each of the six phases of the model in more detail, using existing literature to clarify and provide examples where needed.

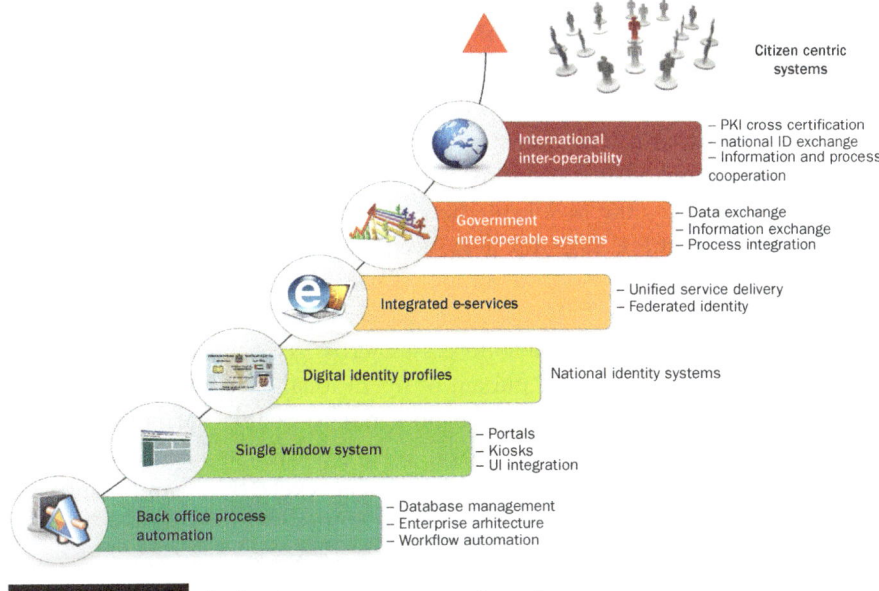

Figure 4.17 A six-stage e-government road map

7.1 Back office process automation

Back office tasks are those that do not require direct interaction with citizens and can be performed more efficiently and effectively off-site. Major reforms have taken place in the public sector administration in Arab countries in the past ten years, and more intensively after the global financial crisis. Arab Governments have shown interest and have been striving to improve the efficiency and effectiveness of their operations. Central to these reforms has been the establishment of governance practices, including the application of modern technology in delivering information and transaction services to citizens.

Arab Governments have shifted many of the front desk processes in the public sector to be performed by back office operators. They have succeeded to some degree in introducing systems that automate and streamline back office business processes under the umbrella of e-government. However, existing back office operations in Arab countries are very paper and labour-intensive. Thus, many of the government services still require citizens to visit different departments to access public services, even after the introduction of ICTs, as systems are not interconnected (UNDP, 2007).

The primary reason for this is that Arab countries have focused on process automation rather than carrying out process re-engineering (transformation) prior to automation, i.e. the same old procedures put into automated forms. This is considered, in our opinion, to be one of the factors that has hindered e-government progress in Arab countries.

Arab governments need to pay more attention to re-engineering initiatives that take a holistic view of the government and attempt to create a business case for enhanced cost-to-income ratios, increase operational efficiencies, and gain a greater share of citizen satisfaction. The same initiatives need to focus on gathering undocumented knowledge and standardising repeatable processes across government agencies. This should be the first step in pursuit of operational excellence in the public sector. Such initiatives have the potential to enable harmonisation of customer service levels across government agencies, and bring significant productivity gains, i.e. reduced processing cycle times and improved quality, visibility and transparency of operations.

7.2 Single window system to deliver public services

In a single window system, access to public services is realised through a governmental portal. Most governments in Arab countries have their own national governmental portals. However, despite the efforts and investments made to date, those portals are still in the first and second phases of Layne and Lee's (2001) e-government evolution model (Al-Khouri and Bal, 2008). Based on our knowledge and interaction with e-government initiatives in the Arab world, we tend to argue that the majority of e-government in Arab countries fail and are stuck in the access phase of Forrester's maturity model. The other evolved Arab countries with e-government are still in the early steps of the interaction phase.

Success in the delivery of public services depends on numerous factors such as flexibility, accessibility, completeness, easiness, security and so on. A single window system that delivers public services and combines all these factors is among the hot topics in government business today. A single window system or 'one-stop window-shopping' concept refers to the integration of public services from a customer of public services' point of view (Wimmer and Tambouris, 2002). It allows citizens, businesses and other authorities to have 24-hour access to public services from their home, their office or even on the move using different access media and devices.

A significant trait of a single window system is accessibility, which should enable multiple delivery and interaction channels between citizens and the government. Each of the channels may have its uniqueness and its value added in terms of speed, convenience, and in the way it allows citizens to interact and communicate with the government at designated offices, call centres, or through the Internet, mobile devices, cable TV, and so on.

A single window system should allow citizens to have 24-hour access to electronic public services from their homes, libraries, schools, shopping malls or even on the move. Figure 4.18 provides an overview of the various service channels and their applicability and general effectiveness.

Implementing a strategic plan is always a big challenge and a channel strategy is no exception. In other words, developing a single window system with multiple delivery channels is a complex and multi-faceted domain that is not only associated with technological systems but also organisational, legal and social. Providing services with a citizen-oriented view calls for a new, service-oriented design approach (e-Europe, 2002).

For an effective citizen-oriented system, we suggest that service providers in the public sector need to adopt hybrid models that enable citizens to interact with the government using multiple channels. Such a multi-channel strategy can address objectives of today's public bodies akin to improving the services provided to the user community and/or reducing the costs of providing its services (European Commission, 2004).

In hybrid service models, a communication established through one channel should be available from another channel to provide a seamless user experience. This also allows unified modes of service delivery and service fulfilment to provide enhanced service options for the citizens to choose from.

As the services and delivery channels mature, service delivery capability maturity becomes established. As service maturity improves, responsibility and accountability in public service become apparent and the government as a whole can stand to guarantee the availability of the delivered services. Service level agreements could be openly published and the efficiency of public service measurements publicly available.

A government that can guarantee such service level agreements for its citizens would then be a true citizen-inclusive e-government. But the journey has only just begun; there is a long way to go before citizens are able to interact seamlessly and derive the true benefits of e-government. This seamless experience is possible when a unified identity is established for the citizen, as the next section elaborates.

Service type	Service category	Service scope	Service availability	ID requirements	Service channels	Service owner	Service recipient
1 The counter	Informational	Service request submission		Not Required		Service Provider	General public
	Transactional	Service delivery	8hr x 5d	Strong ID proof required	OTC		Registerd user
	Commercial	Service fulfillment		ID required with signature for proof of delivery			Identified user
2 On-Site-Over the machine	Informational	Service request submission	16hr x 5d	Not mandatory, but good to identify	Self service Kiosks		General public
	Transactional	Service delivery	16hr x 5d	Strong ID proof required	Assisted kiosks		Registerd user
	Commercial	Service fulfillment	8hr x 5d	ID required with Digital signature with time stamp			Identified user
3 OFF-Site-Over the machine	Informational	Service request submission		Not mandatory, but good to identify	Self service kiosks		General public
	Transactional	Service delivery	16hr x 7d	Strong ID proof required	Assisted kiosks		Registerd user
	Commercial	Service fulfillment		ID required with signature with time stamp			Identified user
4 Remote-Over the internet	Informational	Service request submission		Strong ID proof required	Web browser		General public
	Transactional	Service delivery	24hr x 7d	Strong ID proof required	Smart phone		Registerd user
	Commercial	Service fulfillment		ID required with signature with time stamp			Identified user
5 Remote-Over the phone networks	Informational	Service request submission	24hr x 7d	ID validation required	Smart phone		General public
	Transactional	Service delivery	NA	-	Iphone		Registerd user
	Commercial	Service fulfillment	NA	-	Androids		Identified user

Figure 4.18 Service types and channels

7.3 *Digital identity profiles*

In order to build trust in e-government, we have emphasised in our previous articles the need to develop digital identity management systems to provide services such as user identification, authentication, and authorisation in an e-government environment (Al-Khouri, 2012). A digital identity management system is based on a schema for representing digital identities (a database subset, for example, that includes name, last name, date of birth, photo, certificate, serial number, and so on), and authentication mechanisms and protocols that entities use to demonstrate they are the owners of a given digital identity (Windley, 2004). Accordingly, the purpose of a digital identity is to tie a particular transaction or a set of data in an information system to an identifiable individual. With the help of a digital identity, a user can be identified and authorised to use a given resource or service (Corradini et al., 2007).

Identities and identity management are of primary importance for governments as they encompass the identification of citizens and their interactions with public services and government institutions (Mont, 2002). Trusted, secure and accountable identity management solutions are key e-government enablers. Current systems used in e-government schemes generally do not address security needs, as they merely rely on fixed information to authenticate a user whose full identity might be revealed.

Many Arab countries have initiated advanced national identity management programs in the last ten years, for example the UAE, Oman, Bahrain, Qatar, and Saudi Arabia. These initiatives are based on sophisticated technologies such as biometrics, smart cards and a public key infrastructure that, together, provide strong government-issued digital identity profiles for their population.

The digital identities in these countries are packaged in different ways based on individual national policies. For instance, the national identity card of the UAE has rolled fingerprint biometrics, digital certificate and a unique identity number that make up the digital identity of the card holder. Other countries have either one or two of these three parameters with different biometric systems such as iris and facial recognition. One impact of the digital identity profile is on the types of services that can be delivered across different channels.

A trusted, secure and accountable identity management system is a principal facilitator for governments to go up the ladder of e-government maturity. Governments need to understand that identity management is about the management of digital identity and profile information. Governments need to deploy and promote widespread adoption of an open, flexible, policy-driven, context-aware identity management system that reaches across multiple-service contexts; it needs to be integrated with other management aspects including authentication, authorisation, provisioning and data consolidation, along with related trust, security and privacy aspects (Corradini et al., 2007).

7.4 Integrated e-services: e-government delivery frameworks

Multi-channel integration opens up opportunities for moving more processes to self-service channels, reducing costs for the banks and increasing access for citizens. In fact, it presents opportunities for government agencies to standardise and automate business processes that contribute unequivocal value.

Integration, as a more abstract concept, could mean to bring some parts together and make them a coherent whole (Goldkuhl, 2008). Such integration could mean that different information systems are integrated into one system. Integration could, however, also mean that the parts remain as separate entities but that they work together in a well-functioning manner as federated systems (ibid).

Many countries worldwide have struggled with e-government and service delivery frameworks. The subject of the right enterprise architecture models has been controversial. In Arab countries, deployed customer relationship models vary from one country to another in terms of access channels, execution, delivery and customer service criteria. The UAE, for example, has developed the e-Services Delivery Excellence Model that provides mandatory standards and optional best practices that guide government agencies in Dubai on how to develop and deliver e-services, as well as how to evolve toward a culture of e-government excellence (DEG, 2009).

This phase in our model focuses on areas such as usability and ease of service delivery process, performance, reliability, connectivity and, most importantly, security. While end-users may not encounter these issues directly, they still contribute to e-service access, delivery and execution and, together, will provide a means of seamless and user-centred service delivery.

Rabaiah and Vandijck (2009) proposed a comprehensive e-government framework that defines some strategic building blocks to develop e-government and citizen services. The framework, which has been modularised for flexibility, extensibility and customisability, incorporates very important components of the front office and back office views (see Figure 4.19). The entire e-government development is based on a strong ICT approach with interconnected government departments and integrated databases offering single window services to the citizens over multiple channels. This is the essence of e-government.

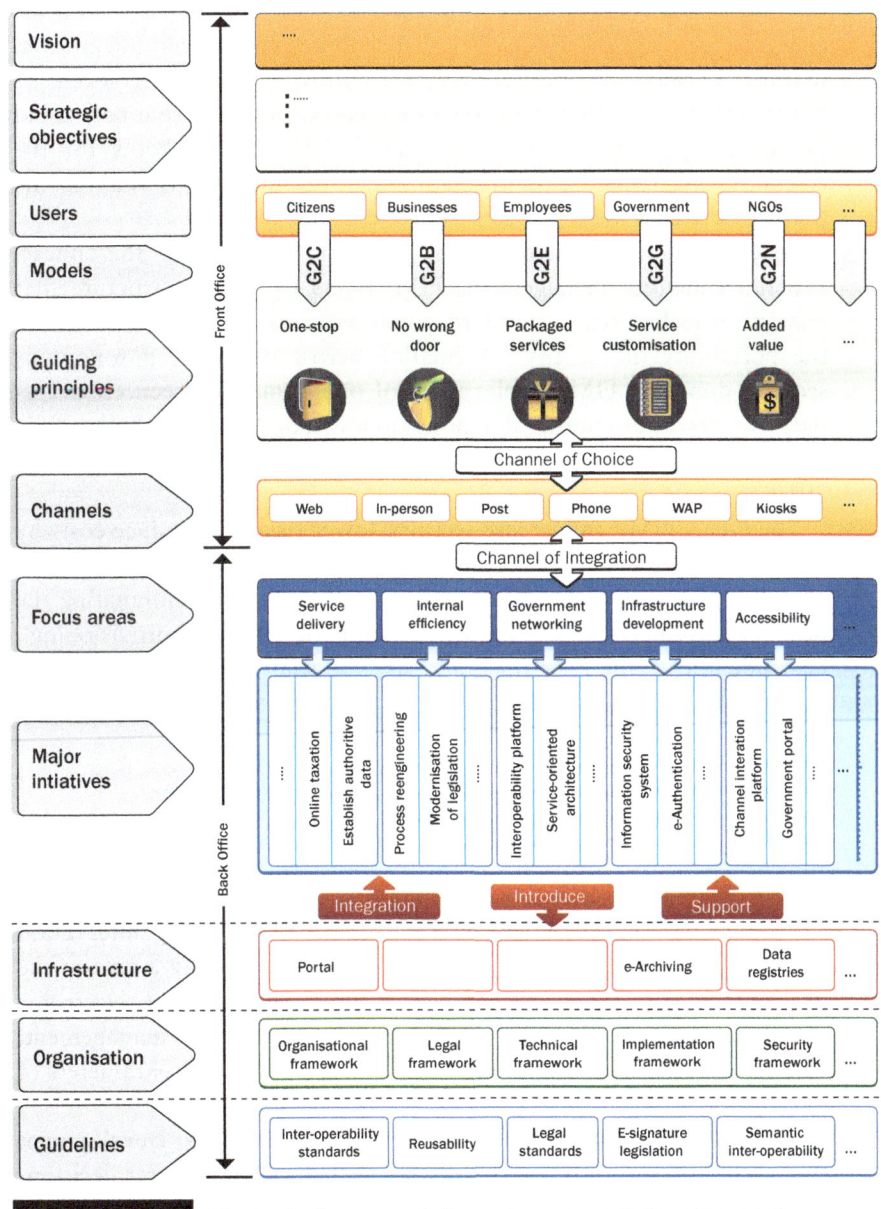

Figure 4.19 Strategic framework for e-government development

Rabaiah and Vandijck's (2009) framework interestingly suggests channel integration for a unified service delivery mechanism. e-Government initiatives are primarily focused on the citizen and the enablement of the inclusion of the citizen in the government. At the basic level, one would look at providing information over multiple channels that citizens can choose from to interact with the government. With the evolution of e-government practices and enhanced back-end integration, maturity in the delivery of services over unified channels is achieved. As citizen satisfaction with the services increases, the same channels could mature into popular and transparent decision-making tools for the citizens' voice to be heard through the government, thus shaping the policies of the nation.

Federated identity management plays a key role in this phase. Federated identity management (FIM) refers to an infrastructure that consists of technologies, standards and user-cases which serve to enable the portability of identity information across otherwise autonomous security domains. FIM enables users of one domain to securely access data or systems of another domain seamlessly, and without the need for a completely redundant user administration (Bertino and Takahashi, 2010).

The use of FIM can increase security, lower risks, and reduce costs by eliminating the need to deploy multiple identity management systems. It can also drastically improve the end-user experience by eliminating the need for account registration through automatic 'federated provisioning', i.e. authenticate the user once and use the same identity information across multiple systems (also referred to as single sign-on).

7.5 Government inter-operable systems: an inter-operability framework

As the Arab governments forge ahead in their quest for an evolved e-government, they need to keep an eye on the evolving integration models. This is critical for government agencies that play a role in the development of public services. The inter-operability of electronic services and communication protocols, unified identity management, and business process integration form some of the various parameters of inter-operability.

Inter-operability, as defined by the United Nations Development Program, refers to 'the creation of systems that facilitate better decision-making, better coordination of government agency programmes and services in order to provide enhanced services to citizens and businesses, the foundation of a citizen-centric society, and the one-stop delivery of services through a variety of channels' (UNDP, 2007).

It is imperative that, within each country, e-initiatives are able to benefit the entire rather than a dispersed narrow section served by isolated departments. In the UAE, for example, the seven Emirates have their own e-services and service channels. Each of the Emirates provides its own e-government portal that enables the local population to interact with the governments. Until recently, these initiatives were confined to each of the Emirates.

Standardisation and inter-operability is being sought now and work is afoot to ensure the seamless integration of services across the seven Emirates. The UAE's government identity management infrastructure is envisaged to play a key role to standardise how people will access information and be authenticated. The application of a federated identity management system, combined with a PKI and national population certification authority (CA),* is expected to bring about a paradigm shift in the citizen experience in how they are introduced to e-services.

Identity management integration is a major contributor to inter-operability. A complete inter-operability model would have to be based on more consideration of data exchange, information exchange, communication protocols, network inter-operability, and process integration.

The definition of inter-operability, put forth by the European Commission's European Inter-operability Framework, is seen as very relevant in the Arab bloc. It defines inter-operability, within the context

---

* In cryptography, certificate authority, or certification authority (CA) is an entity that issues digital certificates. The digital certificate certifies the ownership of a public key by the named subject of the certificate. This allows others (relying parties) to rely upon signatures or assertions made by the private key that corresponds to the public key that is certified. In this model of trust relationships, a CA is a trusted third party that is trusted by both the subject (owner) of the certificate and the party relying upon the certificate. CAs are characteristic of many public key infrastructure (PKI) schemes. Governments worldwide have started owning their own CAs specifically as part of their identity management systems, such national ID card me, electronic passports and so on.

of European public service delivery, as 'the ability of disparate and diverse organisations to interact towards mutually beneficial and agreed common goals, involving the sharing of information and knowledge between the organisations, through the business processes they support, by means of the exchange of data between their respective ICT systems' (European Union, 2011).

Seen in this context of inter-operability, and keeping the regional sensitivities and socio-political environment development of intra and inter-government inter-operability, is seen more as a need than a mere idea.

However, repeated failures to build working systems show that the task is not only difficult but also poorly understood (Navakouski and Louis, 2012). Navakouski and Louis (2012) proposed a framework for understanding inter-operability in the e-government context and how to address its requirements and challenges from policy-making and system development perspectives (see Figure 4.20).

The proposed framework has three primary components: technical, semantic, and organisational. These components are governed by legal, political and socio-cultural factors. At the basic level, inter-operability should enable data exchange at the technical level, where systems can exchange data securely. These are essentially database integration, web services, SOA protocols, and depend on network connectivity.

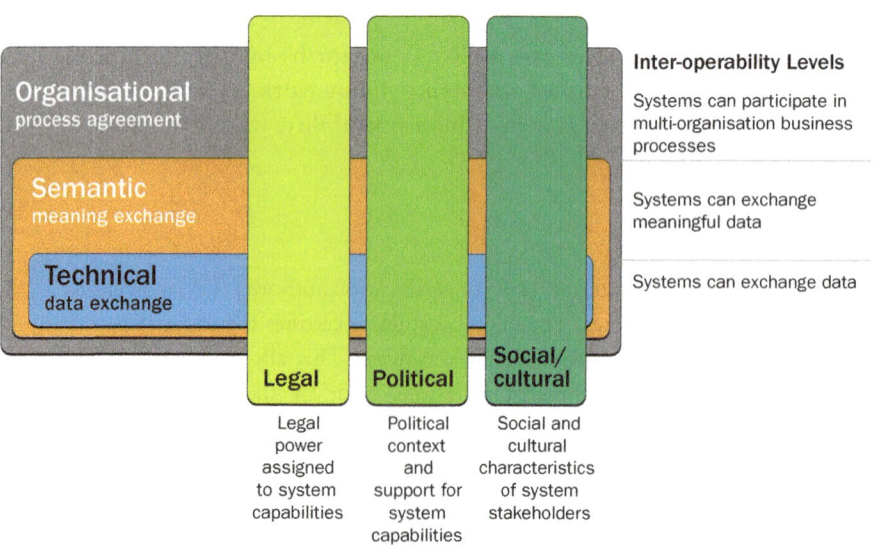

Figure 4.20   e-Government inter-operability model

At the next level is the exchange of information that enables content exchange in terms of the meaning of the information. At the third level is the business process integration and agreements on processes in different departments to basically agree with and complement each other. This handles the policies and procedures to ensure that there is no cross-dependency or counter-dependencies. The last thing one would want is two government departments working at loggerheads! This counter-dependency is even more relevant in the Arab bloc countries that share common borders and common cultural sensitivities.

7.6 International inter-operability

Inter-operability issues also have an international facet, including different levels of conformance and implementation strategies across countries and regions. Even in a world where the international community cooperates to minimise inter-operability problems, parallel ongoing development activities in Asia, Europe, and America will inevitably lead to inter-operability issues (Gallagher and Jeffrey, 2006).

The e-Government Inter-operability Framework (e-GIF) introduced by the UK government in 2004 covers communication, not just within government, but the exchange of information between government systems and citizens, intermediaries, businesses (worldwide), and other governments (UK/EC, UK/US, and so on). The framework covers technical policies and specifications for achieving inter-operability and ICT systems coherence across the public sector. It defines the essential pre-requisites for joined-up and web-enabled government and aims to reduce the costs and risk of operating information technology systems, while keeping the public sector in step with the global Internet revolution.

Another framework is the European Inter-operability Framework (EIF) adopted by the European Union (European Commission, 2010). It provides a basis by which to move on effectively with the implementation and realisation of inter-operability in the area of e-government services and to move on to more and better public services in Europe.

The Framework sets a number of general principles, which should be considered for any e-government service to be set up at a pan-European level: accessibility, multi-lingualism, security, privacy, subsidiarity, use of open standards, assessing the benefits of Open Source Software, and use of multilateral (or 'many-to-many') solutions. Based on these principles, the EIF addresses five distinct levels that need to be considered for the development of cross-border and pan-European e-government services (see Figure 4.21).

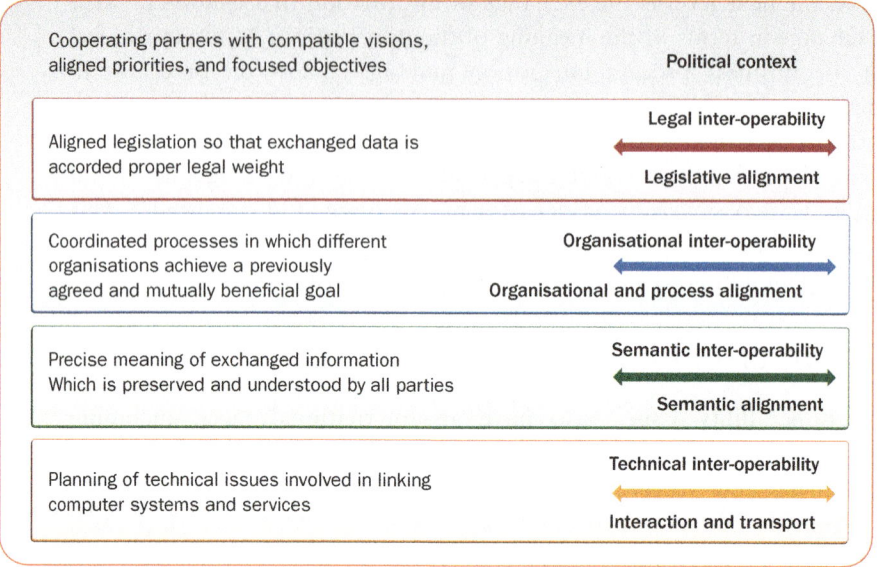

Figure 4.21 European inter-operability Framework Levels (European Commission, 2004)

- Political inter-operability is concerned with political support and sponsorship to provide better, more user-focused public services and cross-border inter-operability efforts to facilitate cooperation among public administrations.

- Legal inter-operability contributes to the provision of a European legal framework that is related to the exchange and protection of data between member states to provide European public services validity across borders.

- Organisational inter-operability is concerned with the definition of business goals and processes by different administrations working together to deliver a service.

- Semantic inter-operability refers to the possibility of the precise meaning of exchanged information to be understandable by any application not initially developed for this purpose.

- Technical inter-operability covers the technical issues of linking-up computer systems, including key areas such as open interfaces, middleware, accessibility and security services.

Arab governments need to agree on and follow similar frameworks that should facilitate business across national borders. To meet this objective, Arab governments need to design and maintain their own version of e-GIF that can take into consideration cultural, political and other technical contexts. This should ensure that the adopted e-GIF provides Arab government agencies with a supportive framework that is aligned with the international environment of inter-operability policies, standards and guidelines. We emphasise that the Arab e-GIF framework needs to be based on the use and adoption of internationally accepted standards. As such, bespoke policies, standards and guidelines will only be developed where deemed strictly necessary. Wherever feasible and relevant, the e-GIF should:

- Utilise existing information and technology policies, standards and guidelines to enable the seamless flow of information across government/public service organisations.
- Mirror established and open international standards for inter-operability.
- Draw upon the inter-operability framework developed in other jurisdictions, e.g. the EIF and UK e-GIF.
- Set practical standards using stable and well-supported products.
- Provide support, guidance and toolkits to enable the standards to be met.
- Provide a long-term strategy that is able to accommodate and adapt.

Global inter-operability requires the architecture to support operations between different systems that represent and handle digital identity using different formats and schemas and have no previously-established trust relationship (Safelayer Labs, 2012).

Arab countries need to pay attention to the inter-operability of identity management systems between governments, which may be seen as a desirable requirement to prevent fraud and theft, e.g. in the case of a biometric passport which needs to be readable and holders' identities verified at the departure and arrival ports of two different international airports (EU JRC, 2005). Inter-operability needs to evolve towards higher levels of flexibility and capability to react to global changes. Their functionalities need to be orchestrated with other management aspects, including trust, privacy and security management.

GCC* countries have initiated multiple projects related to advanced applications development of their e-identity schemes. GCC countries are currently working on developing a unified e-identity infrastructure to enable the identification and authentication of GCC citizens in any of the GCC member states (Al-Khouri and Bachlaghem, 2011). This is argued to raise serious inter-operability issues due to the different and complex infrastructure set ups in each member state and would likely challenge such a project (ibid).

All in all, there is a growing recognition worldwide that inter-operability is essential to the information society (Miller, 2004) and to deliver on the promise of government transformation (Pardo and Burke, 2008). In an increasingly interdependent global economy, the task of developing inter-operability standards for practical applications becomes more complicated, but the rewards become even greater (Miller, 2004). Inter-operability and the use of data for management, decision-making, and governance can be powerful forces for development and prosperity across the globe, or they can effectively divide countries, people, and markets into independent and separate groups (ibid).

Inter-operability within a country or between Arab countries requires leadership and authority. Regardless of context, local, national, or international, inter-operability is an important foundational capability for government transformation (Pardo and Burke, 2008). A more inter-operable government system can change the nature of democracy and citizen participation, and provide systems for service improvement, efficient and effective government operations, and the development of stable and vital economies (ibid).

8. Conclusion

'The most important thing in communication is to hear what isn't being said'.

Peter F. Drucker

---

* GCC is the acronym for Gulf Cooperation Council, also referred to as the Cooperation Council for the Arab States of the Gulf (CCASG). It includes six countries, namely Bahrain, Kuwait, Oman, Qatar, Saudi Arabia, and the United Arab Emirates. The GCC population is estimated to comprise around 40 million people.

Our research presents some useful implications for both research and practice. With regard to research, we have attempted to provide a focused view of the existing literature surrounding e-government and its potential role in improving public sector performance. It also provided some deeper insights to support understanding of related practices in the Arab world.

With regard to practice, our study offers implications for policy-makers and public administrators. The proposed road map provides governments with six primary stages as possible determinants of e-government progress in their contexts. Accordingly, more attention may be paid to these stages to enhance e-government maturity in their settings.

By and large, the presented road map in this article was an attempt to provide a high level conceptual set of steps for Arab governments to focus on in order to progress in the field of e-government and transform the public sector. It promotes business process transformation in government back offices rather than blind automation. It also emphasises the provisioning of integrated e-services through a single, one-stop window and pays heed to inter-operability, both at government and regional levels. The proposed road map can help Arab governments to move in the direction of building citizen-centric systems and looks at development priorities from the point of view of the citizens.

Arab governments need to shake up existing knowledge base about e-government. The existing body of knowledge is, to some extent, perplexing where government practices have been shaped and influenced by mere IT suppliers or private sector consultancy firms. Governments need to create the 'intent' to accept the change in mindset that citizens must be in the heart of any development plan. Change should be from within the government and must reflect their practices and deployed systems.

The citizen-centric government should seek to provide higher levels of customer satisfaction. Governments need to treat themselves as service delivery organisations, where citizens become customers and are integrated in the delivery system design and development. Such delivery systems need to be designed to provide citizens with multiple and integrated interfaces and channels to transact with the government through a single window system.

From a technical perspective, providing integrated services can only be realised if all public authorities are interconnected and their systems are inter-operable. Governments need to recognise that cost reduction and quality improvement can only be achieved when processes are redesigned, databases integrated and, perhaps, certain tasks centralised (Kubicek and Cimander, 2005). In many of these cases, IT systems have to be redesigned, and systems which once stood alone have to be integrated and be able to exchange data with other systems (ibid.). They have to become inter-operable.

Arab governments need to have the ability to monitor e-government progress and evolution in their countries. This should support sustaining the momentum of complex, cross-departmental projects, promoting good governance, and ensuring that the change management challenges of government-level projects are overcome successfully. The proposed road map was developed for this specific need.

In short, Arab public administration needs a radical transformation. The modernisation of public administration is a long-term task, requiring tenacity and innovative thinking. Some Arab countries are now embarking on a new phase in the development of e-government. They need to pay more attention to the overall e-government concept in modernising their nations and building stronger economies that can speak the language of the future.

References

1. Al Borgan, Z. (2011) 'Electronic Government Initiatives in the Arab World 2011'. Arab Advisors Group. Available at: *http://www.arabadvisors.com*
2. Al-Khouri, A. M. and Bachlaghem, M. (2011) 'Towards Federated Identity Management Across GCC: A Solution's Framework'. *Global Journal of Strategies & Governance* 4 (1): 30–49.
3. Al-Khouri, A. M. (2012) 'PKI in Government Digital Identity Management Systems'. *European Journal of e-Practice* 14: 4–21.
4. Andersen, K. V. and Henriksen, H. Z. (2006) 'e-Government maturity models: Extension of the Layne and Lee model'. *Government Information Quarterly* 23 (2): 236–248.
5. Baier, T., Zirpins, C. and Lamersdorf, W. (2003) 'Digital identity: How to be someone on the net'. In Isaias, P. and Reis, A. (eds), *Proceedings of the IADIS International Conference on e-Society* 2: 815–820. Lisbon, Portugal: IADIS Press.
6. Baum, C. and Maio, A. D. (2000) 'Gartner's Four Phases of e-Government Model'. Stamford: Gartner Group Inc.
7. Beniger, J. R. (1986) 'The Control Revolution: Technological and Economic Origins of the Information Society'. Cambridge, Mass: Harvard University Press.

8. Bertino, E. and Takahashi, K. (2010) 'Identity Management: Concepts, Technologies, and Systems'. Boston, MA: Artech House Publishers.
9. Bhatnagar, S. (2008) 'Building Trust through e-Government: Leadership and Managerial Issues'. United Nations Public Administration Network. Available at: *http://unpan1.un.org/intradoc/groups/public/documents/unpan/unpan025871.pdf*
10. Birch, D. (2002) 'Public Participation in Local Government: A survey of local authorities'. Local and Regional Government Research Unit, Office of the Deputy Prime Minister: London. Available at: *http://www.communities.gov.uk/documents/localgovernment/pdf/145623.pdf*
11. Bukhsh, F. A. and Weigand, H. (2011) 'e-Government controls in service-oriented auditing perspective: Beyond Single Window'. In Overbeek, S., Tan Y. H. and Zomer, G. (Eds). *Proceedings of the 1st Workshop on IT Innovations Enabling Seamless and Secure Supply Chains, In conjunction with the EGOV 2011*: 76–90. Delft: CEUR.
12. Camp, J. (2003) 'Identity in Digital Government: A report of the 2003 Civic Scenario Workshop'. National Science Foundation and The Kennedy School of Government. Available at: *http://www.ljean.com/files/identity.pdf*
13. CARICOM (2009) 'Improved Government, Better Service: Draft 2010–2014 CARICOM e-Government Strategy'. Caribbean Centre for Development Administration. CARICAD. Available at: *http://www.gov.gd/egov/docs/ict_egov/draft_2010_2014_CARICOM_egovernment_strategy.pdf*
14. Chandler, S. and Emanuel, S. (2002) 'Transformation Not Automation'. In: *Proceedings of 2nd European Conference on e-Government*: 91–102. St. Catherine's College, Oxford, UK.
15. Chen, H. (2002) 'Digital Government: Technologies and Practices'. *Decision Support Systems* 34: 223–227.
16. Clarke, R. (2001) 'Person location and person tracking: Technologies, risks and policy implications'. *Information Technology & People* 14 (2): 206.
17. Corradini, F., Paganelli, E. and Polzonetti, A. (2007) 'The e-Government digital credentials'. *International Journal of Electronic Governance* 1 (1): 17–37.
18. DEG (2009) 'eServices Delivery Excellence Model'. Dubai e-Government Department. Available at: *http://www.deg.gov.ae/SiteCollectionDocuments/Content/English/eServices%20Delivery%20Excellence%20Model.pdf*
19. Deloitte and Touche (2001) 'The Citizen as Customer'. *CMA Management* 74 (10): 58.
20. Di Maio, A., Kreizman, G., Harris, R.G., Rust, B. and Sood, R. (2005) 'Government in 2020: Taking the Long View'. Garnter Research Group. Available at: *http://www.gartner.com/resources/136400/13646-6/government_in_2020_taking_th_136466.pdf*
21. Dutta, S. and Bilbao-Osorio, B. (eds) (2012) 'The Global Information Technology Report 2012: Living in a Hyperconnected World'. Available at: *http://www3.weforum.org/ docs/Global_IT_Report_2012.pdf*
22. e-Europe (2002) 'Web-based Survey on Electronic Public Services'. Available at: *http:// europa.eu.int/information_society/eeurope/egovconf/documents/pdf/eeurope.pdf*

23. EU JRC (2005) 'Biometrics at the frontiers: Assessing the impact on society'. European Commission Joint Research Centre, Seville, Spain, Tech. Rep. EUR 21585 EN.

24. European Commission (2004) 'Multi-channel delivery of e-Government services'. Available at: *http://www.cisco.com/web/DE/pdfs/publicsector/ida_07_04.pdf*

25. European Commission (2004) 'The European Interoperability Framework'. Final EIF Version 1.0 available at: *http://ec.europa.eu/isa/documents/isa_annex_ii_eif_en.pdf*

26. European Commission (2010) 'Communication Towards interoperability for European public services (COM(2010) 744)'. Available at: *http://eur-lex.europa.eu/LexUriServ/LexUriServ.do?uri=COM:2010:-0744:FIN:EN:PDF*

27. European Commission (2010) 'i2010 – A European Information Society for growth and employment'. Available at: *http://ec.europa.eu/information_society/eeurope/i2010/index_en.htm*

28. European Communities (2004) 'European Interoperability Framework for Pan-European e-Government Services'. Luxembourg: Office for Official Publications of the European Communities. Available at: *http://europa.eu.int/idabc*

29. European Union (2011) 'European Interoperability Framework (EIF): Towards Interoperability for European Public Services'. Luxembourg: Publications Office of the European Union. Available at: *http://ec.europa.eu/isa/documents/eif_brochure_2011.pdf*

30. European Union (2012) 'e-Government in Sweden'. *e-Practice*. Available at: *http://www.epractice.eu/fil-es/eGovernmentSweden.pdf*

31. Feather, J. P. (2008) 'The Information Society: A Study of Continuity and Change'. London: Facet Publishing.

32. Fernandez, A. and Oviedo, E. (ed) (2011) 'e-Health in Latin America and the Caribbean: progress and challenges'. *United Nations publication*. Available at: *http://www.eclac.org/publicaciones/xml/0/44450/2011-205-eHealth_in_LAC_WEB.pdf*

33. Roussos, G., Peterson, D. and Patel, U. (2003) 'Mobile Identity Management: An Enacted View'. *International Journal of Electronic Commerce* 8: 81–100.

34. Gallagher, M. D. and Jeffrey, W. A. (2006) 'Technical and Economic Assessment of Internet Protocol Version 6 (IPV6)'. National Telecommunications and Information Administration and the National Institute of Standards and Technology. Available at: *http://www.ntia.doc.gov/legacy/ntiahome/ntiageneral/ipv6/final/IPv6final.pdf*

35. Gibson, P. D., Lacy, D. P. and Dougherty, M. J. (2005) 'Improving Performance and Accountability in Local Government with Citizen Participation'. *The Innovation Journal: The Public Sector Innovation Journal* 10 (1): 1–12. Available at: *http://www. innovation.cc/volumes-issues/gibson1.pdf*

36. Goldkuhl, G. (2008) 'The challenges of Interoperability in e-government: Towards a conceptual refinement'. Pre-ICIS 2008 SIG e-Government Workshop, Paris. Available at: *http://www.vits.org/publikationer/dokument/664.pdf*

37. Goldkuhl, G. (2012) 'From policy to design and effects: A framework for e-government research'. 9th Scandinavian Workshop on e-Government, February 9–10, 2012, Copenhagen. Available at: *http://www.vits.org/publikationer/dokument/766.pdf*

38. Hai, J. C. (2007) 'Fundamental of Development Administration'. Selangor: Scholar Press.

39. Hayden, C. T., Sanford, A. L., McGivern, D. O., Cohen, S. B., Dawson, J. C., Bennett, R. M., Johnson, R. M., Bottar, A. S., Tisch, M. H., Farley, E. L., Chapey, G. D., Gardner, A. B., Frank, C. K., Philips, H., Bowman, J. E. and Cores-Vazquez, L.A. (2002) 'Participation in Government'. The University of the State of New York. Available at: *www.p12.nysed.gov/ciai/socst/documents/partgov.pdf*

40. Hilty, L. M. (2011) 'Localisation technologies Project'. University of Zurich. Available at: *http:// www.ta-swiss.ch/?uid=146*

41. Homburg, V. (2008) 'Understanding e-Government: Information Systems in Public Administration'. London: Routledge.

42. Howard, M. (2001) 'e-Government across the Globe: How will "e" Change Government'? *Government Finance Review* 17 (4): 6–9.

43. Ifinedo, P. and Singh, M. (2011) 'Determinants of e-Government Maturity in the Transition. Economies of Central and Eastern Europe'. *Electronic Journal of e-Government* 9 (2): 166–182.

44. Stoddart, J. (2006) 'PIPEDA review discussion document: Protecting privacy in an intrusive world'. Office of the Privacy Commissioner of Canada, Ottawa.

45. Kearney, A. T. (2011) 'How to Become a Citizen-Centric Government'. Available at: *http://www.atkearney.com/images/global/pdf/Citizen-Centric_Government.pdf*

46. Klewes, J. and Wreschniok, R. (eds) (2009) 'Reputation Capital: Building and Maintaining Trust in the 21st Century'. Heidelberg: Springer.

47. Kubicek, H. and Cimander, R. (2005) 'Interoperability in e-Government. A Survey on Information Needs of Different EU Stakeholders'. Available at: *http://www.ifib.de/publikationsdateien/IOP_in_eGov_-_Survey_on_Information_Needs.pdf*

48. Layne, K. and Lee, J. (2001) 'Developing Fully Functional e-Government: A Four Stage Model'. *Government Information Quarterly* 18 (2): 122–136.

49. Lee, T., Hon, C. T. and Cheung, D. (2009) 'XML Schema Design and Management for e-Government Data Interoperability'. *Electronic Journal of e-Government* 7 (4): 381–390.

50. Lenihan, D. G. (2008) 'Realigning Governance: From E-Government to E-Democracy'. In Anttiroiko, A. (ed). *Electronic Government: Concepts, Methodologies, Tools, and Applications*: 3389–3422. Hershey PA: IGI Global.

51. Lowery, L. M. (2001) 'Developing a Successful e-Government Strategy'. United Nations Public Administration Network. Available at: *http://unpan1.un.org/intradoc/groups/ public/documents/APCITY/UNPAN000300.pdf*

52. Lowndes, V., Pratchett, L. and Stoker, G. (2001) 'Trends in public participation: part 1 – local government perspectives'. *Public Administration* 79 (1): 205–222.

53. Lozano, L., Hilbert, M., Takahashi, T., Legale, E. and Virapatirin, M. (2003) 'Basic Document: Network 13 Towns & The Information Society'. Available at: *http://centrourbal.com/sicat2/documentos/07_2009723448_R13P3-04A-db1-eng.pdf*

54. Mattelart, A. (2003) 'The Information Society: An Introduction'. London: Sage Publications Ltd.

55. Merriam-Webster Dictionary (2011). Available at: *http://www.merriam-webster.com/dictionary/government*

56. Miller, R. B. (2004) 'Toward Global Interoperability'. *Directions Magazine* April 30. Available at: *http://www.directionsmag.com/articles/toward-global-interoperability/123759*

57. Monga, A. (2008) 'e-Government in India: Opportunities and challenges'. *Journal of Administration & Governance* 3(2). Available at: *http://joaag.com/uploads/5_ Monga2EGov3_2_.pdf*

58. Mont, M. C., Bramhall, P., Gittler, M., Pato, J. and Rees, O. (2002) 'Identity Management: a Key e-Business Enabler'. Hewlett-Packard Laboratories, UK. Available at: *http://www.hpl.hp.com/techreports/2002/HPL-2002-164.pdf*

59. Navakouski, M. and Louis, G. (2012) 'Interoperability in the e-Government Context'. Software Engineering Institute, Carnegie Mellon University. Available at: *http://www.sei.cmu.edu/reports/11tn014.pdf*

60. Nordfors, L., Ericson, B. and Lindell, H. (2006) 'The Future of e-Government: Scenarios 2016'. VINNOVA – Swedish Governmental Agency for Innovation Systems. Available at: *http://www.vinnova.se/upload/EPiStorePDF/vr-06-11.pdf*

61. OECD (2008) 'Making Life Easy for Citizens and Businesses in Portugal: Administrative Simplification and e-Government'. Organisation for Economic Co-operation and Development. Available at: *http://www.oecd.org/dataoecd/37/23/42600869.pdf*

62. OECD (2009) 'Government at a Glance 2009'. Organisation for Economic Co-operation and Development. Available at: *http://www.oecd-ilibrary.org/docserver/download/fulltext/4209151e.pdf?expires=1341518042&id=id&accname=guest&checksu m=7F5931A65265D9C0C63B69986E84D79D*

63. OeE (2004) 'e-Government Interoperability Framework'. Version 6.0, Office of the e-Envoy (OeE). Available at: *http://edina.ac.uk/projects/interoperability/e-gif-v6-0.pdf*

64. Otenyo, E. E. and Lind, N. S. (2011) 'e-Government: The Use of Information and Communication Technologies in Administration'. Youngstown, New York: Teneo Press.

65. Pardo, T. A. and Burke, G. B. (2008) 'Government Worth Having: A briefing on interoperability for government leaders'. Center for Technology in Government, University of Albany, SUNY. Available at: *http://www.ctg.albany.edu/publications/reports/government_worth_having/government_worth_having.pdf*

66. Rabaiah, A. and Vandijck, E. (2009) 'A Strategic Framework of e-Government: Generic and Best Practice'. *Electronic Journal of E-Government* 7 (3): 241–258.

67. Rahav, A. (2012) 'Citizen Centricity'. Icentred blog. Available at: *http://www.icentered.com/citizen-centricity*

68. Safelayer Labs (2012) 'Defining an architecture for the complete management of digital identity'. Available at: *http://labs.safelayer.com/en/research-and-development/focus-areas/iam-30/326-definicion-de-una-arquitectura-para-la-gestion-integral- de-la-identidad-digital*

69. Seifert, J. W. and McLoughlin, G. J. (2007) 'State e-Government Strategies: Identifying Best Practices and Applications'. A report prepared for the Congressional Research Service by the Lyndon Baines Johnson School of Public Policy at the University of Texas at Austin. Available at: *http://www.fas.org/sgp/crs/secrecy/RL34104.pdf*

70. Sherry, D., Ryan, S. D., Zhang, X., Prybutok, V. R. and Sharp, J. H. (2012) 'Leadership and Knowledge Management in an e-Government Environment'. *Administrative Sciences* 2: 63–81.

71. Song, H. J. (2006) 'e-Government in Developing Countries: Lessons Learned from Republic of Korea'. United Nations Educational, Scientific and Cultural Organisation (UNESCO). Available at: *http://www2.unescobkk.org/elib/publications/083/e-government.pdf*

72. Suh, S. Y. (2007) 'Promoting Citizen Participation e-Government: From the Korean Experience in e-Participation'. Available at: *http://unpan1.un.org/intradoc/groups/ public/documents/un/unpan020076.pdf*

73. Suthrum, P. and Phillips, J. (2003) 'Citizen Centricity: e-Governance in Andhra Pradesh'. The University of Michigan Business School. Available at: *http://www.bus.umich.edu/FacultyResearch/ResearchCenters/ ProgramsPartnerships/IT-Champions/ eGovernance.pdf*

74. UNPAN (2010) 'United Nations e-Government Global Report's'. UN Public Administration Programme. Available at: *http://www2.unpan.org/egovkb/global_ reports/10report.htm*

75. UNDP (2007) 'e-Government Interoperability: Guide'. United Nations Development Programme. Available at: *http://www.ibm.com/ibm/governmentalprograms/undp-gif-guide.pdf*

76. UNDP (2007) 'e-Government Interoperability'. Overview, United Nations Development Programme, 2007. Available at: *http://www.apdip.net/projects/ gif/GIFOverview.pdf*

77. United Nations Department of Economic and Social Affairs (2010) 'United Nations e-Government Survey 2010'. UN. Available at: *http://www2.unpan.org/egovkb/documents/2010/E_Gov_2010_Complete.pdf*

78. United Nations (2012) 'United Nations e-Government for the People'. United Nations Department of Economic and Social Affairs. Available at: *http://unpan1.un.org/intradoc/groups/public/documents/un/unpan048065.pdf*

79. West, D. M. (2004) 'e-Government and the Transformation of Service Delivery and Citizen Attitudes'. *Public Administration Review* 64 (1): 15–27.

80. Wimmer, M. A. and Tambouris, E. (2002) 'Online One-Stop Government: A working framework and requirements'. In *Proceedings of the IFIP World Computer Congress*, August 26–30, Montreal.

81. Windley, P. J. (2004) 'Digital Identity: Unmasking Identity Management Architecture'. Sebastopol, CA: O'Reilly.

82. World Bank (2012) 'Definition of e-Government'. World Bank. Available at: *www.worldbank.org/egov*

83. World Economic Forum (2010) 'The Lisbon Review 2010: Towards a More Competitive Europe'? Available at: *http://www3.weforum.org/docs/WEF_LisbonReview_Report_2010.pdf*

e-Voting in the UAE FNC elections: a case study

Abstract: Electronic voting (e-voting) has been attracting the attention of governments around the world. Many countries have pursued implementing e-voting systems in their national elections. This article presents a case study of an e-voting system deployment in the United Arab Emirates (UAE). The UAE has conducted its Federal National Council (FNC) elections for the 2011–2015 session of the National Assembly using an advanced e-voting system with biometric-based smart cards to verify voters' identities. The article provides detailed insights on the phases of the project, from the design phase up to election day. The article also provides a comprehensive overview of the current literature around e-voting in order to enhance understanding of the field and of global practices.

Keywords: *Electronic voting, UAE FNC elections, national identity.*

1. Introduction

Electronic voting is gaining in popularity around the world. It has become well-established in countries such as Belgium and Switzerland, and has been deployed in many European countries such as the UK, Denmark, France, and Ireland. Electronic voting has been considered to be an efficient and cost-effective alternative compared to the traditional classic voting procedures (IPI, 2001). See also Figure 5.1. Electronic voting technologies that have been in use vary from punch cards, optical scan voting systems and specialised voting kiosks, including self-contained direct-recording electronic (DRE) voting systems. Additional technological components may also enable the transmission of ballots and votes via telephones, personal handheld computer devices, and the Internet.

Figure 5.1 Conventional ballot-based voting system

The purpose of this article is to provide a more thorough analysis of a government e-voting system deployment. The case study provides an overview of the UAE Federal National Council (FNC) elections, which were based on an advanced e-voting system. The voter-recognition system in the UAE FNC elections was based on a smart card with biometrics and public key cryptography.

The structure of this paper is as follows. In Section 2, we describe the methodology adopted. Section 3 presents an introduction to the concept of elections and voting, and the different approaches to voting. Section 4 provides an overview of the existing literature around e-voting. Section 5 provides a brief glance at e-voting system deployment around the world. Section 6 presents the UAE FNC elections case study, outlining the different phases of the project, from the design phase up to election day. Section 7 presents some key factors that contributed to the overall success of the elections. In Section 8 we outline some lessons learnt from the implementation of the UAE FNC e-voting system. Finally, we draw some conclusions and provide some pointers for future work.

2. Research methodology

The research content presented in this article aims to provide an understanding of the issues surrounding the use of electronic voting in practice. It thus provides a review of the existing literature in the field of e-voting to outline some of the critical points in the current body of knowledge. The researcher uses an action-based case study method to provide a contextual analysis of a recent government implementation of an e-voting system. Indeed, there is a continuum between the 'describer' of case studies and the 'implementer' of action research (Waddington, 1994).

The case study method was used to describe relationships that exist in reality (Yin, 1984) and to produce an understanding of the context of the information and the process whereby the information system influences and is influenced by the context (Walsham, 1993). The action research aimed to develop outcomes and solutions that are of practical value to its stakeholders, while at the same time developing and contributing to the existing body of knowledge (Rapoport, 1970; Susman and Evered, 1978).

The senior role of the researcher in the reported project as a member of the higher national election committee, and head of the technology team, as well as a member of other management and legal committees, enabled him to have a more in-depth and working view of the case, and acquire understanding of specific situations. In addition, the researcher's fieldwork involved the design of the system and overall observation of its implementation, from the early phases of preparation, design during the run-up to the election and on the actual polling day.

An agile project and system development methodology was adopted, which is discussed in some detail in Section 6.3. Semi-structured workshops with key stakeholders and commercial suppliers' staff were undertaken before implementation to allow for government concerns and requirements to be addressed. The researcher's observatory role on election day took place at the operations management centre, which was set up to handle the technical and organisational issues that arose. The researcher was also part of the verification processing team at the end of election day. All this supported the research work and provided the opportunity to acquire hands-on experience of the implementation, management, and administration of the e-voting system.

3. Elections and voting

An election is a formal decision-making process by which a population chooses an individual(s) to hold a public office position(s) (Britanica, 2012). Elections are associated with the term 'electoral reform', which describes the process of introducing fair electoral systems, or improving the fairness or effectiveness of existing systems that should altogether reflect the public opinion.

Electronic voting is a term that may encompass several different types of voting, embracing both electronic means of casting a vote and electronic means of counting votes. Electronic voting systems were first debuted when punched card systems were introduced for the 1964 presidential elections [Saltman, 1975]. Then, optical scan voting systems emerged that allowed computer systems to count voters' marks on ballots, i.e. direct-recording electronic (DRE) voting machines.

DRE voting machines collect and tabulate votes in a single machine, and have been used in elections in Brazil and India, and also on a large scale in Venezuela and the United States. They have also been used on a large scale in the Netherlands, but have been decommissioned after public concerns. There has been controversy, especially in the United States, that electronic voting, especially DRE voting, can facilitate electoral fraud.

Nonetheless, the concept of e-voting has been gaining popularity in many countries. Windley (2005) describes the lure of e-voting and the growing applications of digital technologies to voting systems, to the simple idea that computers, and the Internet, have fundamentally changed other parts of our lives; he states that, since voting is one of the basic processes of democracy, it seems a natural candidate for electronic automation.

Advocates of e-voting argue that the use of advanced information technologies not only speeds up the counting of ballots, but also brings about advanced features of uniqueness, accuracy, completeness, verifiability, auditibility, privacy, and uncoercibility.

In general, two main types of e-voting can be identified:

- e-Voting is physically supervised by representatives of governmental or independent electoral authorities (e.g. electronic voting machines located at polling stations); this is the most common and preferred approach (see also Table 5.1 for e-voting systems).

- i-Voting is also referred to as remote e-voting, where voting is performed within the voter's sole influence, and is not physically supervised by representatives of governmental authorities (e.g. voting from one's personal computer, mobile phone, television via the Internet). Internet voting systems have gained popularity and have been used for government elections and referendums in the United Kingdom, Estonia and Switzerland, as well as municipal elections in Canada and party primary elections to use in the United States and France.

There are also hybrid systems that include an electronic ballot marking device (usually a touchscreen system similar to a DRE) or other assistive technology to print a voter-verified paper audit trail, then to use a separate machine for electronic tabulation.

Table 5.2 provides an overview of the typical strengths and weaknesses that different e-voting solutions tend to have compared to paper-based equivalents (Internet voting vs postal voting; voting machine vs paper voting in controlled environments) (IDEA, 2011).

Table 5.1   Types of e-voting systems

Voting system	Description
Paper-based electronic voting system	Also referred to as 'document ballot voting system', first emerged as a system where votes are casted and counted by hand, using paper ballots. With the advent of electronic tabulation came systems where paper cards or sheets could be marked by hand, but counted electronically. These systems included punched card voting, marksense and later digital pen voting systems.
	Most recently, these systems can include an Electronic Ballot Printers (EBPs), that allow voters to make their selections using an electronic input device, usually a touch screen system similar to a DRE machine, that produce a machine-readable paper or electronic token containing the voter's choice. This token is fed into a separate ballot scanner which does the automatic vote count.

Table 5.1   Types of e-voting systems (*Cont'd*)

Voting system	Description
Direct-recording electronic (DRE) voting system	A direct-recording electronic (DRE) voting machine records votes by means of a ballot display provided with mechanical or electro-optical components that can be activated by the voter (typically buttons or a touchscreen); that processes data with computer software; and that records voting data and ballot images in memory components. After the election, it produces a tabulation of the voting data stored in a removable memory component and as printed copy.
	The system may also provide a means for transmitting individual ballots or vote totals to a central location for consolidating and reporting results from precincts at the central location. These systems use a precinct count method that tabulates ballots at the polling place. They typically tabulate ballots as they are cast, and print the results after the close of polling.
Public network DRE voting system (Internet voting systems)	With Internet voting systems, votes are transferred via the Internet to a central counting server. Votes can be casted either from public computers or from voting kiosks in polling stations, or more commonly from any Internet-connected computer accessible to a voter.
OMR Systems	Optical and digital scanning systems which are based on scanners that can recognise the voters' choice on special machine-readable ballot papers. OMR systems can be either central count systems (where ballot papers are scanned and counted in special counting centres) or precinct count optical scanning (PCOS) systems (where scanning and counting happens in the polling station, directly as voters feed their ballot paper into the voting machine).
	These are normally used to improve the accuracy of the counting process and reduce potential manual counting errors. However, the quality of the count depends on the correct marking of the ballot paper and the quality of the ink used by the voter.
Polling stations	At a polling station there is one medium to record the vote, which is then registered in a ballot box on another device. This system differs substantially from a DRE in that nothing is storVVed in the DRE and it is impossible for a voter to manipulate the memory containing the vote.

Table 5.2					

Strengths and weaknesses of e-voting systems

Electoral issues compared to paper voting	Internet voting	DRE with out VVPAT	DRE with VVPAT	PCOS	Electronic ballot printers
Faster count and tabulation	Strength	Strength	Strength	Strength	Strength
More accurate results	Strength	Strength	Strength	Strength	Strength
Management of complicated electoral systems	Strength	Strength	Strength	Strength	Strength
Improved presentation of complicated ballot papers	Mixed	Mixed	Mixed	Weakness	Mixed
Increased convenience for voters	Strength	Mixed	Mixed	Weakness	Mixed
Increased participation and turnout	Strength	Neutral	Neutral	Neutral	Neutral
Addressing needs of a mobile society	Strength	Mixed	Mixed	Neutral	Mixed
Cost savings	Mixed	Weakness	Weakness	Weakness	Weakness
Prevention of fraud in polling station	Neutral	Strength	Strength	Strength	Strength
Greater accessibility	Mixed	Mixed	Mixed	Weakness	Mixed
Multi-language support	Strength	Strength	Strength	Weakness	Strength
Avoidance of spoilt ballot papers	Strength	Strength	Strength	Strength	Strength
Flexibility for changes handling of deadlines	Strength	Strength	Strength	Weakness	Strength
Prevention of family voting	Strength	Neutral	Neutral	Neutral	Neutral

Table 5.2 Strengths and weaknesses of e-voting systems (*Cont'd*)

Electoral issues compared to paper voting	Internet voting	DRE with out VVPAT	DRE with VVPAT	PCOS	Electronic ballot printers
Lack of transparency	Weakness	Weakness	Mixed	Mixed	Mixed
Only experts can fully understand the voting technology	Weakness	Weakness	Mixed	Mixed	Mixed
Secrecy of the vote	Weakness	Mixed	Mixed	Mixed	Mixed
Risk of manipulation by outsiders	Weakness	Mixed	Mixed	Mixed	Mixed
Risk of manipulation by insiders	Weakness	Weakness	Weakness	Weakness	Weakness
Costs of introduction and maintenance	Strength	Weakness	Weakness	Weakness	Weakness
Intrastructure / environmental requirements	Mixed	Weakness	Weakness	Weakness	Weakness
Lack of e-voting standards	Weakness	Weakness	Weakness	Weakness	Weakness
Meaningful recount	Weakness	Weakness	Strength	Strength	Strength
Vendor dependence	Weakness	Weakness	Weakness	Weakness	Weakness
Increased IT security requirements	Weakness	Weakness	Weakness	Weakness	Weakness

Note
Details in the matrix would vary depending on specifics of context and systems. Cases where these details are very important are classified as 'mixed'; cases where e-voting has little or no impact are classified as 'neutral'.
VVPAT: Voter-verified paper audit trail (VVPAT).
PCOS: Precinct count optical scans.

4. Literature review

The amount of reported work on the subject of e-voting is significant. The literature of electronic voting states that the field needs significant improvement (Alexander, 2001; Besselaar and Oostveen, 2003; Cranor, 2000; IPI, 2001; Hargrove, 2004; Liptrott, 2006; Manjoo, 2003a; Manjoo, 2003b; Millar, 2002; O'Donnell, 2002; Oostveen and Besselaar, 2006; Shamos, 1993; Xenakis and Macintosh, 2006). Advocates of e-voting point out that electronic voting can reduce election costs and increase civic participation by making the voting process more convenient. Critics maintain that without a paper trail, recounts are more difficult and may open the door for electronic ballot manipulation, and that even poorly-written programming code could affect election results. Many researchers have produced different studies and approaches to address these concerns. A sample outline of such research is provided in Table 5.3.

A report worthy of note on e-voting was produced by an MIT Universality Team in the United States. The Caltech/MIT Voting Technology Project, which was established in December in 2000 following the controversial election recount of the 2000 presidential vote in Florida, assessed the magnitude of the problems surrounding voting systems, their root causes and how technology can reduce them (Caltech-MIT, 2001). The report provided a set of recommendations on the various issues related to voting and proposed a framework for a new voting system with a decentralised, modular architecture in which vote generation is performed separately from vote casting. The report emphasised the importance of developing a permanent audit trail. It also stressed the fact that the vote generation machine can be proprietary, whereas the vote casting machine must be open-source and thoroughly verified and certified for correctness and security.

The National Institute of Standards and Technology draft report, issued in 2006, proposed vote verification through a parallel process of electronic ballot counts (NIST, 2006). It indicated that voting systems should allow election officials to recount ballots independently from a voting machine's software. The recommendations endorse 'optical-scan' systems in which voters mark paper ballots that are read by a computer, and electronic systems that print a paper summary of each ballot, which voters review and election officials save for recounts. NIST indicated in its report that the lack of a paper trail for each vote 'is one of the main reasons behind continued questions about voting system security and that it diminished public confidence in elections'.

Table 5.3 Research around e-voting systems

Researcher(s)	Contribution
Neumann (1993)	Recommended a list of generic voting criteria to enhance the security of the overall voting systems and resistance to failure. These include confidentiality, integrity, availability, reliability, and assurance for the involved computer systems. He also concluded that, operationally, no commercial system is likely to ever meet all requirements, and that developing a suitable custom system would be extremely difficult and prohibitively expensive.
Mercuri (1993; 2002)	Her philosophy is very similar to Neumann's. She invented her own method for electronic voting; also referred to as the Mercuri method. Her method is similar to the Caltech/MIT proposal where the voting machine in her method is designed to produce human-readable hard-copy paper results, which can be verified by the voter before the vote is casted, and manually recounted later if necessary.
Chaum (2004)	Presented a system design that provides voter verifiability, i.e., to provide voters with the capability to verify that their vote is accurately included in the tally – even if all election computers and records were compromised. Such trust would provide a high degree of transparency that allows close auditing of the vote capture and counting process if needed. Chaum's proposed system design preserves ballot secrecy, while improving access, robustness, and adjunction, all at lower cost.
Shamos (1993)	Provided a sharp counterpoint to Neumann and Mercuri's views, but was less impressed with paper ballots than are Neumann and Mercuri. He recommended DRE machines with a decentralisation design to make fraud difficult to commit and easy to detect. He contribution also involved the development of the 'Six Commandments' summary of requirements for a voting system. The list of requirements can be used to critique voting systems and as basis for public inspection.
Kosmopoulos (2004); Rivest (2004)	Pointed out issues related to the development of secure platforms and issues related to the simplification and usability of voting machines, development of audit-trails, support for disabled voters, security problems of absentee ballots, and so on.
Clausen et al. (2000)	Presented a secure electronic polling system which does not rely on permanent network connections between polling places and the vote-tallying server. They build the system on a disconnected (or, more accurately, an intermittently connected) environment, which works well in the absence of network connectivity.

Table 5.3 Research around e-voting systems(*Cont'd*)

Researcher(s)	Contribution
Volkamer (2009)	Stressed on the need to provide a trustworthy base for secure electronic voting, and how to prevent accidental or malicious abuse of electronic voting in elections. Others refer to the opaque nature of the technologies involved, which few understand, and that it is crucial that electronic voting systems provide a voter-verifiable audit trail, which will act as a permanent record of each vote that can be checked for accuracy by the voter before the vote is submitted, which will make it difficult or impossible to alter after it has been checked (Kosmopoulos, 2004).
Klein (1995)	Presented a remote voting system design to improve the levels of privacy, universal verifiability, convenience and untraceability, but at the expense of receipt-freeness. The suggested design applies the technique of a blinded signature to a voter's ballot so that it is impossible for anyone to trace the ballot back to the voter.

The California Internet Voting Report [CIVTF, 2000] suggested an innovative strategy to enable remote Internet voting to improve participation in the elections process, i.e. providing voters with the ability to cast their ballots at any time from any place via the Internet. On the contrary, experts argue that the Internet is not yet ready for online 'prime time' national federal elections, given the current state of insecurity of hosts and the vulnerability of the Internet to manipulation and denial-of-service attacks (Hisamitsu and Takeda, 2007; Kosmopoulos, 2004; Rubin, 2002; Schneier, 2004). They also identify security issues in social engineering and in using specialised devices and other factors that could undermine the sanctity of an Internet-based election process, and that the current infrastructure is inadequate for remote Internet voting (ibid).

Another report produced by the National Institute of Standards and Technology (NIST) in 2008 concluded that widely-deployed security technologies and procedures could mitigate many of the risks associated with electronic ballot delivery, but that the risks associated with casting ballots over the Internet were more serious and challenging to overcome (NIST, 2008). Another recent NIST report (2011) concluded that Internet voting systems cannot currently be audited with a comparable level of confidence in the audit results as those for polling place systems.

Other researchers point out that building secure online voting systems is far from being possible and that a small configuration or implementation error would undermine the entire voting process (Wolchok et al., 2012). Reference is made to the pilot project of an online voting in Washington, D.C. and how researchers at University of Michigan were able to break through the security functions and gained complete control of the election server in less than 48 hours. Researchers argue that fundamental advances still need to be made in security before e-voting will truly be safe (Cramer et al., 1996; Ikonomopoulos et al., 2002; Schoenmakers, 1999; Wolchok et al., 2012).

Researchers indicated that operationally, no commercial system is likely to ever meet all requirements, and that developing a suitable custom system would be extremely difficult and prohibitively expensive (Neumann, 1993). Others indicated that any catastrophic failures and sweeping fraud made possible by imperfections in electronic voting machines are also likely to occur in a real election (Shamos, 1993).

Shamos (1993) refers to the fact that the real source of election problems is the result of human limitations. He claims that the chief source of the issue is the willingness of unsuccessful politicians to embrace any conceivable reason for their loss, except that the voters did not want them. His reflective recommendation proposes that government efforts expended in meeting threats to the election process should be rationally related both to the probability of the threat and the seriousness of its effects.

Other researchers argue that the progress of e-voting relies on the advancement of standards and technical solutions, which should take into account discussions on general requirements, threat perceptions and the economic, political and sociological implications surrounding the use of electronic voting systems (Alexander, 2001; Cranor, 2000; Hillman, 2007; Hoffman, 2004; Jones, 2001; Shamos, 1993; UK POST, 2001; Volkamer, 2009). Rubin (2002) argues that technologists should take on a role to educate the policy-makers about the issues surrounding an e-voting system and enable them to develop more effective strategies.

Another part of the literature deals with public trust and confidence. It argues that, if the public perceives elections to be unfair, the foundation of the government is weakened. Whether electronic voting systems are fair may not even matter; it is the public perception that is crucial (Bonsor and Strickland, 2011).

On a different standpoint, the International Institute for Democracy and Electoral Assistance published a policy report that identified some essential considerations for e-voting systems to gain public trust and confidence (IDEA, 2011). It introduced a pyramid of trust that consists of three levels: credible electoral process, socio-political context, and operational and technical context (see Figure 5.2).

The top level represents the ultimate goal of electoral reform and the introduction of the e-voting system, and is a factor that is dependent on the two levels shown below. Public trust is seen to be determined by the socio-political context in which e-voting is introduced. Some factors in this context can be directly addressed by a comprehensive e-voting implementation strategy, while others, such as a general lack of trust in the Election Management Body (EMB), or fundamental political or technical opposition, will be more difficult to change. A negative socio-political context has the potential to create serious risks, even if the technical and operational foundations of the e-voting solution are sound.

Chiang (2009) developed a technology acceptance model that constitutes four trust variables, namely ease of use, perceived usefulness, attitude of usage, and security. The research results showed the effect of 'ease of use' on voters' attitude towards using the e-voting system required 'perceived usefulness' as a medium. Then, the effect exhibited positive and significant influences among ease of use, perceived usefulness and attitude towards using the e-voting system. The security of the e-voting system has a positive and significant effect on attitude and trust. The study concluded that the security of the e-voting system plays an important role in establishing user trust. Overall, the literature identified key elements that e-voting systems need to heed. These are listed in Table 5.4.

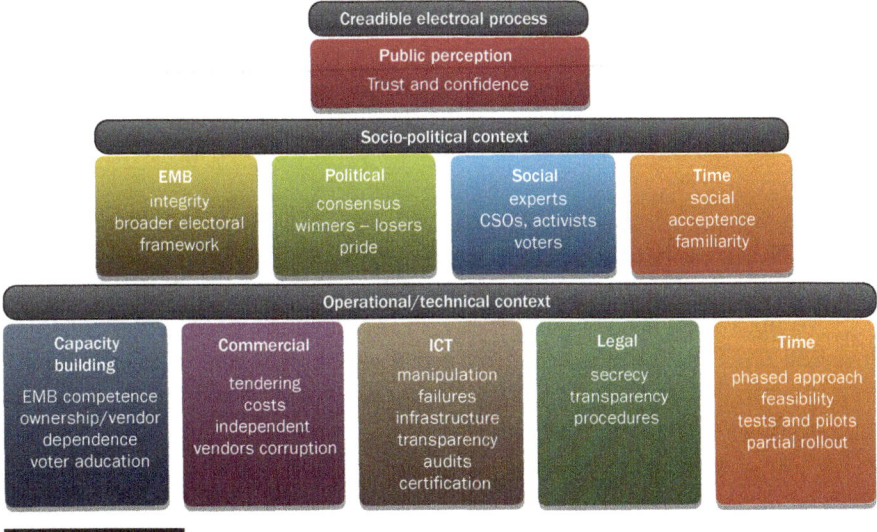

Figure 5.2 The pyramid of trust

Table 5.4 Voting system requirements

Requirement	Description
Privacy	To ensure the secrecy of the ballots.
Universal Verifiability	To ensure that all valid votes have been included in the final tally.
Robustness	The system can tolerate a certain number of faulty participants.
No receipt	The voters cannot provide a 'receipt' that shows what they voted.
Fairness	No partial tally is revealed before the end of the elections.
No disputes	The fact that the participants follow the protocol at any phase can be publicly verified by any casual third party.
Self-tallying	The post-ballot-phase can be performed by any interested third party.
Ballot Secrecy	To ensure the secrecy of what takes place. The only thing revealed about the voters' choice is the final result.
Authentication	To ensure that individuals cannot be impersonated
Accuracy	To ensure that each individual's vote is recorded and counted
Timelines	To record information and to have the results available quickly
Accessibility	To have a system that is accessible to all and easy to use
Security	To guard against manipulation and interference.

Though debate on the issue of e-voting has been and will continue to be passionate, most critics recognise that a move towards an electronic voting system is an inevitable step in the evolution of the voting process (2007). Governments seem to be motivated to adopt e-voting systems, despite the issues and concerns reported in the literature, as the next section will present.

5. e-Voting deployment around the world

'No one would buy a safe that could easily be opened, but everyone who has ever bought a safe has bought one that can be cracked. The same is true for voting systems. The issue is not whether they are secure, but whether they present barriers sufficiently formidable to give us confidence in the integrity of our elections'. (Shamos, 1993)

Despite what the literature raises in terms of risks and the reasons why not to go for electronic voting systems, governments worldwide seem to be motivated to implement these systems. The following world map depicts the adoption of the e-voting systems worldwide, as well as some examples of e-voting trials and uses.

As we can see from the map in Figure 5.3, though countries in the West have initiated e-voting, developing countries have taken a noticeable lead in adopting e-voting systems at a national level for their election systems. In fact, the adoption of e-voting seems to be closely associated with maturity

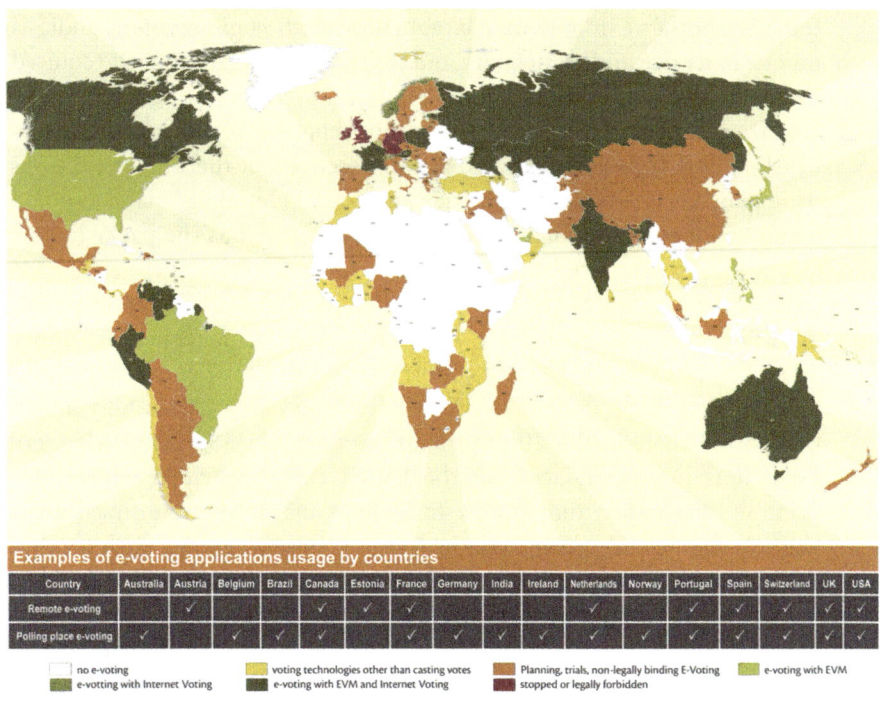

Examples of e-voting applications usage by countries

Country	Australia	Austria	Belgium	Brazil	Canada	Estonia	France	Germany	India	Ireland	Netherlands	Norway	Portugal	Spain	Switzerland	UK	USA
Remote e-voting		✓			✓	✓	✓				✓		✓	✓	✓	✓	✓
Polling place e-voting	✓		✓	✓	✓		✓	✓	✓	✓	✓	✓	✓		✓	✓	✓

- no e-voting
- e-votting with Internet Voting
- voting technologies other than casting votes
- e-voting with EVM and Internet Voting
- Planning, trials, non-legally binding E-Voting
- stopped or legally forbidden
- e-voting with EVM

Figure 5.3 e-Voting around the world (source e-Voting.cc)

in election processes worldwide. The following sub-sections will highlight the use of electronic voting systems in Australia, India and Estonia.

5.1 Australia

Australia has the distinction of compulsory voting at Federal and State levels. Every enrolled eligible voter must vote. Failure to vote can lead to legal implications and fines. Driven by this need of compulsory voting, Australia first used remote voting by electronic means in 2001. In the subsequent elections of 2004, 2007 and 2010, widespread usage of technology ensured the successful deployment of electronic voting. The desire to include all sections of eligible voters, backed by the compulsion to vote, made Australia adopt i-voting (assisted voting) for visually impaired voters. The i-voting system is also available for Australians who live overseas. As a result, Australia is now firmly entrenched in the world map of e-voting as a pioneer in adopting e-voting systems.

5.2 India

India is another example of a large nation with eligible voters. India is unique in its size and complexity and the sheer logistical support required to conduct nation-wide elections. It follows precinct voting, as in many countries. There are nearly 740 million eligible voters spread across nearly 830,000 polling stations. Certain sections of the Indian elections were plagued by voting malpractice and fraud. Handling, managing and manning nearly one million polling stations, as well as the paper ballots and ballot boxes, posed severe security problems.

India first adopted electronic voting machines (EVMs) in 1982, allowing voters to cast their votes in electronic machines in about 50 polling stations. Indeed, the country has come a long way in deploying EVMs. Since 2004, Indian elections have been using EVMs fully and it is reported that, in the 2009 general elections, 1,368,430 EVMs were used. It is now mandated for the EVMs to be provided with ballot printouts for paper audit trails. In 2011, Gujarat was the first Indian State to adopt and use Internet voting for State elections, with around 26 million eligible voters.

The success of Indian electronic voting systems can be gauged from the fact that, today, countries like Bhutan, Nepal, Kenya, and Fiji are set to use these EVMs, with India exporting the machines, technology and resources. While Australia continues to use EVMs in precinct voting, India is moving to a hybrid mode of EVMs and i-voting. Large-scale urbanisation, Internet availability and accessibility and encryption technologies are making it possible for India to begin utilising i-voting.

5.3 Estonia

Estonia is a classic example of a country using technology in conducting national elections using i-voting. Central to their theme was the issuance of an ID card with PKI (digital certificate) capabilities for voter identification, authentication and, finally, casting an encrypted vote ensuring privacy and voter secrecy combined with anonymity.

The digital certificate in the ID card in Estonia is envisaged to enable the voting system to more robustly verify the credentials of the card holder. The certificate itself is submitted to the voting system through the card using the cardholder's PIN. It is then verified for authenticity and validity. Once validated and authenticated, the user is provided access to the on-line voting site. The voter selects the candidate and submits the vote. The vote submission is akin to the voter sealing the paper ballot in a secure envelope and dropping it into the ballot box. The digital certificate is used for digitally signing the vote and encrypting, using the private key of the card holder and the public key of the e-voting system. The vote is submitted as if sealed in a secure envelope. At the server end, decryption of the vote takes place using the election keys, while the signature validity of the voter is checked. At no point is the voter's identity revealed.

Overall, the key success factors for e-voting and i-voting systems in these countries are considered to be due to the existence of a legal framework and legal validity of the use of these systems, and the availability of an audit trail of the counted votes. The latter point enables the voter to go back to the voting system to check and confirm if his/her vote was indeed considered in the counting of the votes.

From our reading of the existing practices, it is interesting to note the factors that have driven countries to adopt e-voting systems. Each country is driven by different factors that were key in their adoption of the e-voting systems. Australia is driven by its need to ensure 'inclusive voting', that is, every eligible voter should be able to vote and should be provided with the access and ability to vote. This includes overseas Australians, visually impaired citizens and so on. This is as a result of its compulsory voting' law.

India is driven by its need to contain election fraud and to reduce the complexity in the logistics of handling manual ballots. Estonia is driven by its need for providing citizen convenience and to increase voter participation. The elections conducted in Estonia were beset with problems of low voter turn-out.

The US has moved to machine-based (EVM) e-voting primarily because of the problems it faced with the punch card voting systems in 2000. The Florida elections were found to have discarded votes, which otherwise were eligible votes and, had they been counted, the election results would have seen a different conclusion.

France, Germany, and Belgium have basically taken advantage of their home-grown technologies and adopted e-voting systems. Brazil had similar issues as in India, but the adoption of DRE (Direct Recording Electronic Voting) Machines were used which were different from the EVMs used in India. Thus, our understanding is that local political needs and socio-economic factors have contributed to the adoption of e-voting systems in different countries.

The purpose of this section was to provide a short overview of some successful implementations of electronic voting systems globally. The next section will present the main content of this article, namely the use of an e-voting system in the UAE FNC elections.

6. The UAE FNC elections

On 2 December 1971, with the adoption of the Constitution, the Federation of the United Arab Emirates was officially established. A few months later, in February 1972, the country's first ever Federal National Council (FNC) was set up as the country's legislative and constitutional body. The FNC consisted of forty members appointed by the rulers of each of the seven Emirates.

In 2005, the UAE had its first national elections. The presidential resolution stipulated that half of the FNC members (out of 40 members) would be elected by citizens and the other half would be appointed by the ruler of each Emirate. This was recognised as a step forward to enhance a well-structured political participation in line with citizens' aspirations, and as a major milestone towards modernisation and development of the Federation.

Indeed, the introduction of this partial election system was seen as the first step in a gradual process aimed at empowering and enhancing the role of the FNC, and developing more effective and vital channels for coordinating between the FNC and the government, thereby opening new prospects in the parliamentary life of the UAE. In the first electoral experience in the UAE in 2005, the NEC approved electronic voting instead of traditional voting procedures.

The same election model was used for the 2011 FNC elections, except for the electoral college, where the number of voters increased from around 6,000 to almost 130,000. The 2011 FNC elections were considered to be more challenging due to the short time frame and the size of the electoral college, as well as the fact that the majority of voters were first-time voters and had never seen a ballot box (see also Figure 5.4). The government decided to take innovative steps to encourage participation and introduced technology-driven systems to facilitate the overall program.

6.1 Forming a National Election Commission (NEC)

Organisation is an important element of the overall management process. It is next to planning in importance. Organising involves the integration of resources in order to accomplish objectives. In management terms, organisation is both the process and the end-product of that process, which is referred to as the organisation structure. The success of the management process will be determined by the soundness of the organisation structure. Therefore, the structure acts as the foundation on which the whole super-structure of management is built.

Clearly, a sound organisation structure was fundamental to ensure that election management activities are conducted in an efficient and effective manner. A National Election Commission (NEC) was formed in February 2011 (seven months before the elections) by a presidential decree consisting of ten government officials representing key government organisations and entities that were envisaged to support the efforts of

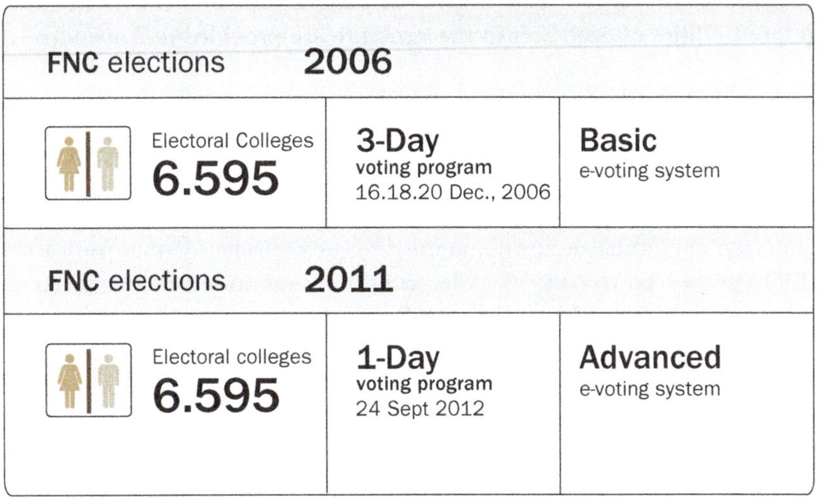

FNC elections	2006		
Electoral Colleges **6.595**	**3-Day** voting program 16.18.20 Dec., 2006	**Basic** e-voting system
FNC elections	2011		
Electoral colleges **6.595**	**1-Day** voting program 24 Sept 2012	**Advanced** e-voting system

Figure 5.4   2006 vs. 2011 FNC elections

regulating the electoral process in addition to three public figures. The NEC was empowered to oversee the whole election process, including:

- Setting out the overall election framework.
- Supervising the elections.
- Supporting efforts to raise electoral awareness.
- Developing election guidelines.
- Locating polling centres in each Emirate.
- Approving regulatory measures for establishing the electoral legal framework.
- Issuing and seeking approval for the governing rules for the lists of members of the electoral panel.
- Setting the date of elections.

Current research states that defining roles and responsibilities within the e-voting system implementation could provide a better understanding of who is responsible for doing what at different stages so that the planned election result is produced (Xenakis and Macintosh, 2006). As such, the commission developed an organisation structure with 20 sub-units to support the election program. With a flat hierarchical organisation structure, the National Election Commission (higher committee) was responsible for establishing strategy and overall direction, whilst the lower level units had the responsibility for a specific function, as illustrated in Figure 5.5. More details about the functions and responsibilities of each unit in the structure are provided in Appendix 1.

6.2 Opting for an e-voting system

The National Election Commission opted for introducing an advanced electronic voting system for the 2011 elections. The national identity management infrastructure maintained by the Emirates Identity Authority (EIDA) was seen to construct the primary foundation for the desired e-voting system and enhance its overall security, in other words the use of the Population Register to extract the electoral roll of eligible voters, and the utilisation of biometric-based smart identity card issued to citizens (e-ID cards) to authenticate the voters.

It was also decided that the e-voting system would be developed with a decentralised architecture. This was because of concerns over the permanence and stability of the e-voting system, lack of reliability on the internet communications, and to ensure continuity and simultaneous

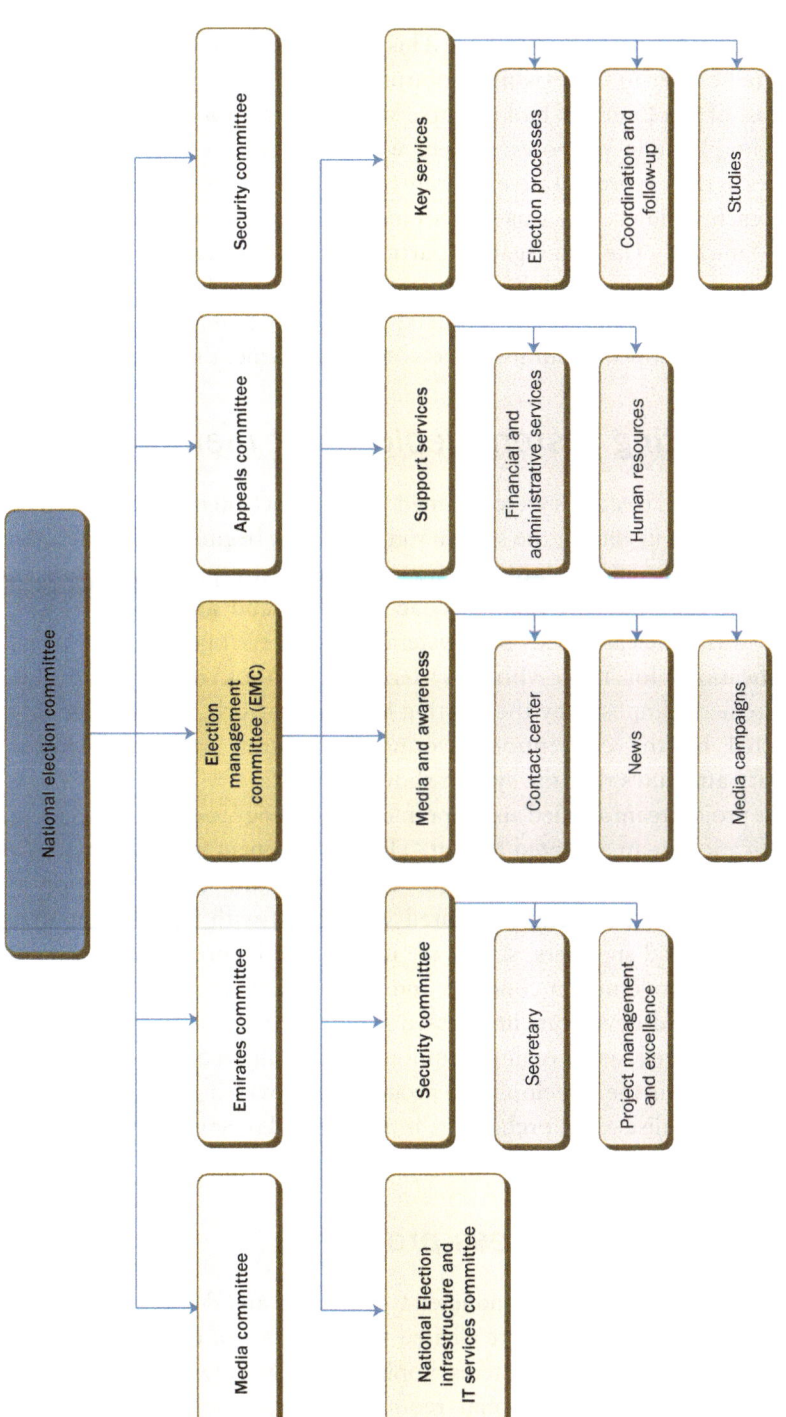

Figure 5.5 National Election Commission structure

operations of the electoral process. This was in line with recommendations reported earlier in the existing literature.

Most of the Emirates had multiple voting centres for contingency and capacity-planning purposes. Voters were expected to visit the polling centres in their representative Emirate to cast their votes. The information from each voting centre was synchronised with the voting centres within that Emirate. The aggregated participation information from each Emirate was sent manually to the Abu Dhabi headquarters, but the ballots of each Emirate were not synchronised (see Figure 5.6). The following section will outline the e-voting deployment methodology.

6.3 e-Voting system deployment methodology

The UAE FNC elections were planned to take place on only one day. The challenge was in getting the system right from the beginning. Besides, the new electronic voting process was envisaged to promote integrity, accuracy, time savings, reliability, accessibility, and auditability. Thus, there was a clear need for systematically developed requirements specifications for the e-voting system, which took into account the requirements imposed by the existing legal framework, the functionality reflected by the conventional voting procedures, and the required security attributes that the system should exhibit.

The project team applied an adaptable methodology to elicit requirement specification in an accepted format. This development methodology was adopted for the design, development, deployment and implementation of the e-voting system. Semi-structured workshops with key stakeholders and commercial suppliers' staff were undertaken before implementation to allow for government concerns and requirements to be addressed.

The adopted development stressed rapid adaptations to the e-voting system, as well as evolving requirements facilitated by direct user involvement in the development process. It provided a framework by which to visualise scale, orchestrate mundane and repetitive development tasks, and enforce processes. (see Figure 5.7).

6.4 e-Voting business processes

Multiple workshops were conducted to articulate and document business processes. Requirements were elicited through a set of cases, along with specifications of similar systems implemented worldwide. These were later translated into functional requirements for the e-voting system design. See also Table 5.5 for the adopted design principles.

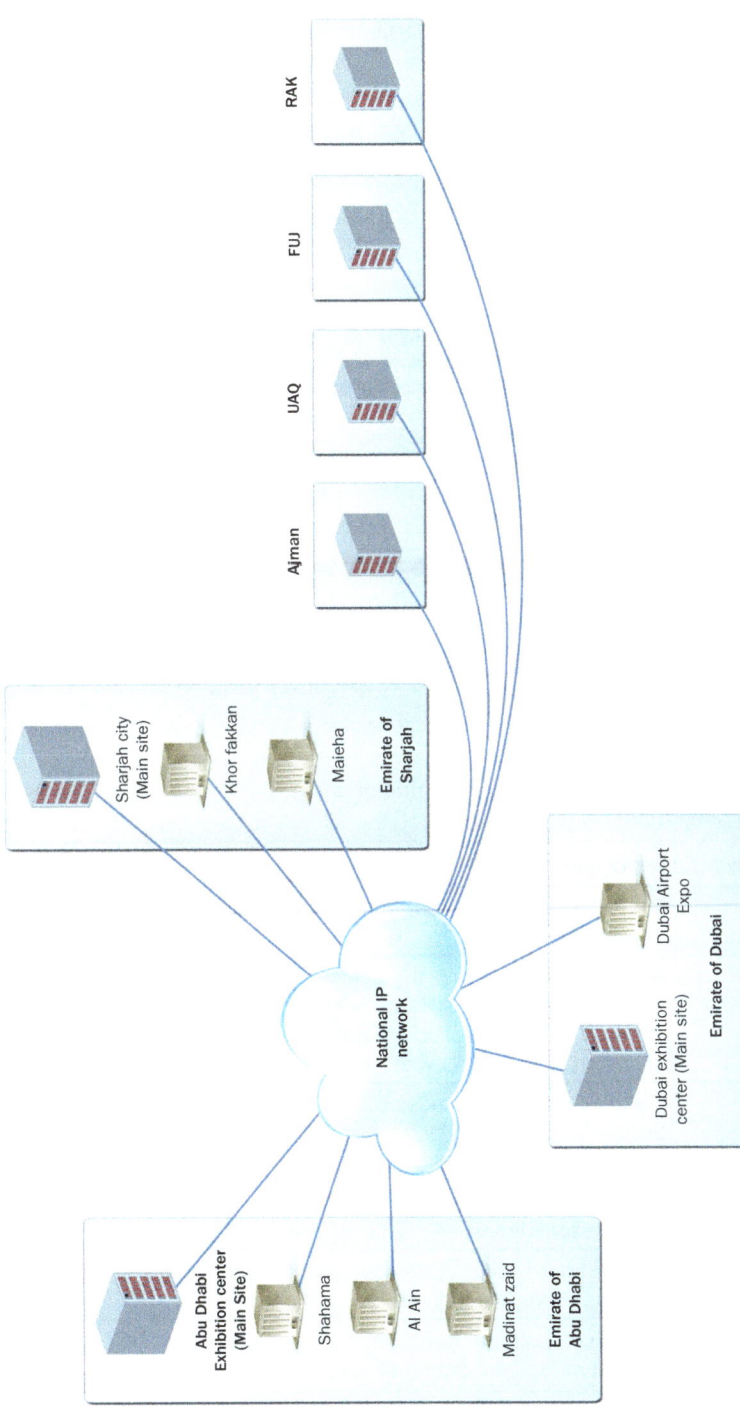

Figure 5.6 FNC e-voting system architecture

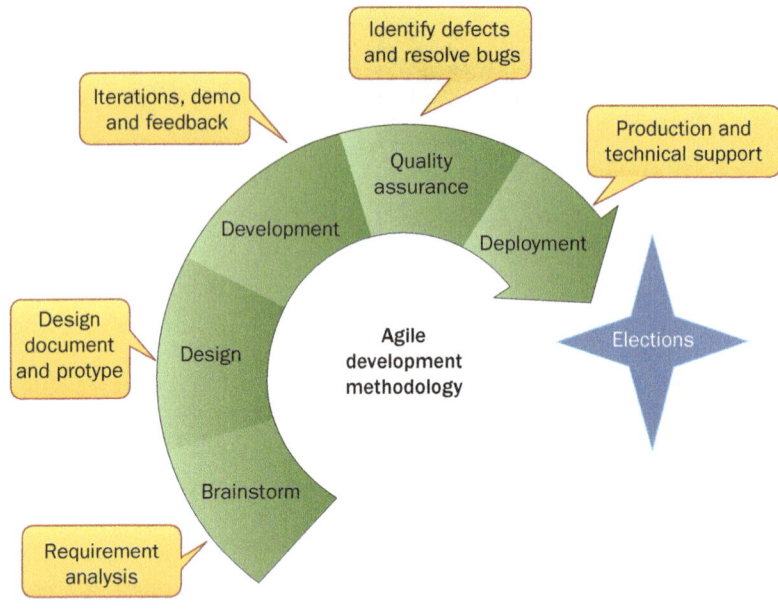

Figure 5.7 The UAE e-voting deployment methodology

Table 5.5 e-Voting design principles

Objective	Description
Voter authenticity	The voting platform must ensure only eligible voters are allowed to vote.
Voter anonymity	Voter privacy regarding the selected voting options must be protected at all levels. Nobody – including the system administrators and election authorities – should be able to associate a voter's ID with his selected voting options.
Election integrity	The election system must ensure only one ballot is cast per eligible voter, and ballots are processed and counted as defined.
Service availability	The voting platform must be available during the prescribed election time limits.
Open auditing	It must be possible to demonstrate the accuracy, integrity, and fairness of the election process.
Service protection	All the components from the voting services must be resistant to system failures, denial of service and security attacks.

The primary input for business processes was taken from the UAE Election Legal Framework that was reviewed and approved by the Federal National Election Commission. It outlined the overall electoral processes. In general terms, the following were identified as the main requirements for the e-voting system:

- Authentication of the voter will rely on the UAE e-ID card.

- De-centralised voting features will be implemented in each Emirate, which will be synchronised with the central database.

- A voting sequence should allow a selection of candidates up to the number of candidates valid for each Emirate.

- A ballot copy of each vote cast will be printed for audit purposes.

- Each ballot copy will contain: election banner, Emirate, voting centre, candidate(s) selection (Number and Name). See also Figure 5.8.

On the other hand, the business processes related to the e-voting system were categorised in three different stages, as outlined in Figure 5.9 and explained below.

6.4.1 Pre-election period

1. The electoral college represented the selected electors (also referred to as 'electoral roll of eligible voters') who would vote for the candidates in their representative Emirates. Each of the seven Emirates had its own electoral college. The electoral college list was extracted from the population register database maintained by EIDA.* An application was developed and deployed in all the seven Emirates, where each Ruler Court selected its representative electoral college.

---

*The Emirates Identity Authority is a federal government organisation in the UAE established in 2004 to develop and maintain a national identity management infrastructure. Its role involves enrolling all the population in the program by 2013 and issuing them with advanced smart identity cards with their biometrics stored in the chip. The Authority is envisaged to become the primary reference for population demographics data to support strategic planning and decision-making in the country. The Authority's role also involves setting up a national validation gateway to support e-government and e-commerce, which all identity verification and authentication transactions will need to go through before the service is availed. Advanced capabilities are possible through UAE national identity cards, like multi-factor authentication (biometrics, pin, digital certificates), encryption and digital signatures.

Figure 5.8   Approved design of the FNC ballot paper

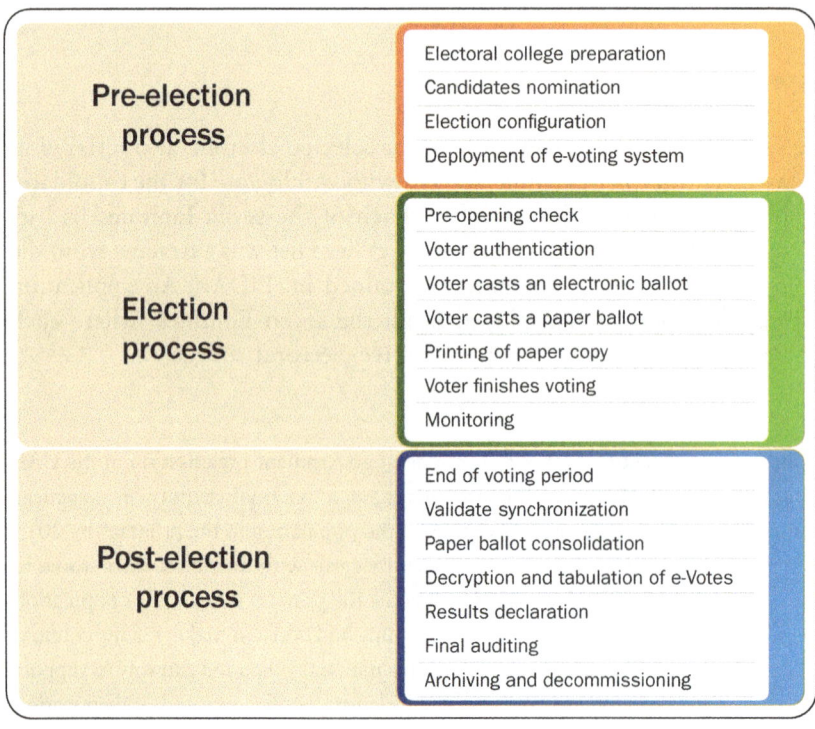

Pre-election process
- Electoral college preparation
- Candidates nomination
- Election configuration
- Deployment of e-voting system

Election process
- Pre-opening check
- Voter authentication
- Voter casts an electronic ballot
- Voter casts a paper ballot
- Printing of paper copy
- Voter finishes voting
- Monitoring

Post-election process
- End of voting period
- Validate synchronization
- Paper ballot consolidation
- Decryption and tabulation of e-Votes
- Results declaration
- Final auditing
- Archiving and decommissioning

Figure 5.9   e-Voting business process categories

2. Candidate registration: the candidates came from the same electoral college list. A clear, standardised, equitable and transparent registration candidate and application review process was essential to maintain electoral integrity and to ensure that each candidate had an understanding of the requirements and were able to register if qualified. Timely notification of acceptance or rejection, and the right to an appeal if required.

6.4.2 Election day

3. Voter verification was carried out using the state-of-the-art smart ID cards issued by the Identity Authority to all UAE citizens. Voter verification was enabled through the identity toolkit developed by the Identity Authority and integrated with the voting system.

4. Electronic voting machines utilising touch screen monitors for casting votes were deployed. Election servers were deployed in all seven Emirates conducting simultaneous polling electronically in 14 centres across the UAE.

5. The ballot box: the e-voting was complemented by a paper print-out of the e-vote to provide a paper trail and verifiability of the electronic vote count.

6.4.3 Post-election

6. The vote count: votes were electronically collected from multiple centres, mixed, tallied and results published within minutes of the closure of the voting.

7. Reports and statistics: from the encrypted databases that provided long-term secure storage, reports were extracted to indicate voter participation, turnout and other vital statistics required by the Government on demand.

A more detailed overview of the finalised business processes implemented in the UAE FNC e-voting system is presented in Appendix 2.

6.5. *Voter volume and flow in the voting centres*

Compared to the UAE Elections in 2005, the number of voters for the 2011 FNC elections was 20 times the previous election; 6000 voters in 2005 and around 130,000 in 2011. This required detailed planning in the areas of site preparation and capacity computation, technical infrastructure development, communications planning, addressing logistical and staff requirements, and the overall specifications of the electronic voting system. The planning phase was fundamental to ensuring effective implementation of the electoral process and to encourage and allow a large number of voters to successfully participate in the election.

First, and based on the expected numbers of voters in each Emirate, NEC identified a number of voting centres to accommodate voter capacity in each and every Emirate. See also Figure 5.10, which depicts the demographics of the FNC electoral college. Next, an analysis using the Erlang model* based on expected voter turnout, average acceptable waiting time, and turn-out expected during peak hours helped determine the number of voter validation desks and voting booths required at each voting centre.

6.5.1 Calculating capacity

Detailed scientific calculations were carried out to determine the exact number of voting terminals required to cater to the voting needs of the electorate. In a zero waiting time scenario, each voter could be provided with an individual ID verification station and a personal voting station, where the voter could walk in anytime, cast the vote and leave. This would

---

*Erlang is a declarative language for programming concurrent and distributed systems, which was developed by the authors at the Ericsson and Ellemtel Computer Science Laboratories. Agner Krarup Erlang (1878–1929) was a Danish mathematician who developed a theory of stochastic processes in statistical equilibrium - his theories are widely used in the telecommunications industry. Erlang's model primitives provide solutions to problems which are commonly encountered when programming large concurrent real-time systems. The module system allows the structuring of very large programs into conceptually manageable units. Error detection mechanisms allow the construction of fault-tolerant software. Code-loading primitives allow code in a running system to be changed without stopping the system. Erlang has a process-based model of concurrency. Concurrency is explicit and the user can precisely control which computations are performed sequentially and which are performed in parallel. A message passing between processes is asynchronous; that is, the sending process continues as soon as a message has been sent. The only method by which Erlang processes can exchange data is message passing. This results in applications which can easily be distributed - an application written for a uniprocessor can easily be changed to run on a multi processor or network of uniprocessors.

Percentage of Women and Men in Electoral Colleges

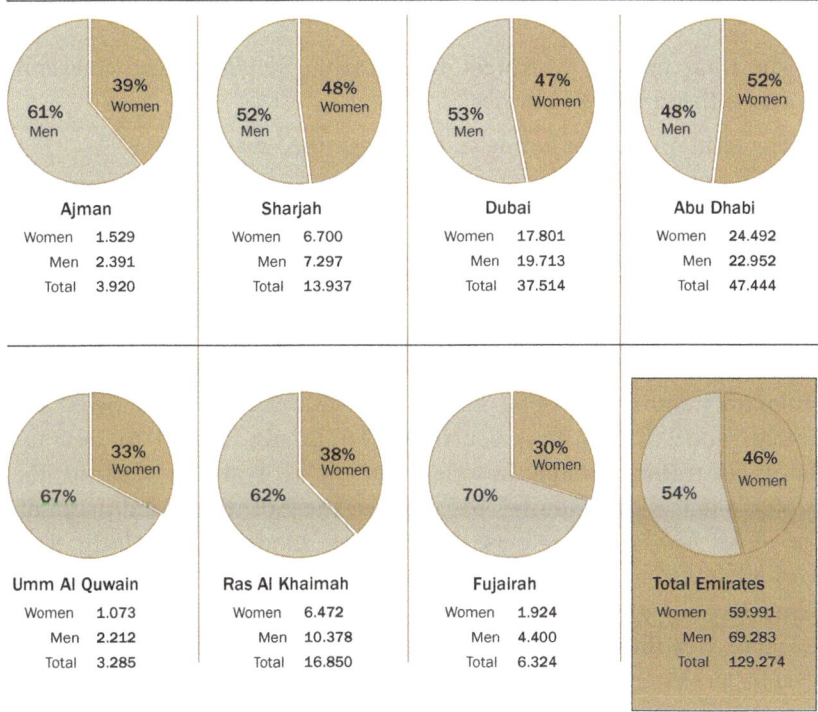

Ajman		Sharjah		Dubai		Abu Dhabi	
Women	1.529	Women	6.700	Women	17.801	Women	24.492
Men	2.391	Men	7.297	Men	19.713	Men	22.952
Total	3.920	Total	13.937	Total	37.514	Total	47.444

Umm Al Quwain		Ras Al Khaimah		Fujairah		Total Emirates	
Women	1.073	Women	6.472	Women	1.924	Women	59.991
Men	2.212	Men	10.378	Men	4.400	Men	69.283
Total	3.285	Total	16.850	Total	6.324	Total	129.274

Women and men in the Electoral colleges (<30 years)

Percentages

	%	
Abu Dhabi	27.58%	
Ajman	37.14%	
Dubai	37.87%	
Fujairah	55.83%	
Ras Al Khaimah	39.7%	
Sharjah	38.44%	
Umm Al Quwain	30.23%	
Total	35.09%	

Electoral colleges

Abu Dhabi	47.444
Ajman	3.920
Dubai	37.514
Fujairah	6.324
Ras Al Khaimah	16.850
Sharjah	13.937
Umm Al Quwain	3.285
Total	129.274

As of 12/07/2011

Figure 5.10   FNC electoral college composition and demographics

have been a very cost-effective way to meet the voting requirements. The capacity planning and calculation was based on the following constraints:

- The total number of voters was nearly 130,000 spread unevenly across the different Emirates.
- The total number of voting stations identified was 13.
- Voting had to be carried out between 8 am and 8 pm.
- Voters had to be verified for their ID.
- Voters had to be given sufficient time to determine their choices at the voting terminal to select their candidates for casting their votes.

This is where the queue modelling and traffic calculations resulted in an optimum number of voting terminals. The calculations followed the Erlang model for traffic calculation. An Erlang is used to describe the total traffic volume in one hour; it is typically used in the telecommunications industry to calculate the call traffic and maintain the SLAs to the subscribers. The Erlang model is popularly used worldwide for traffic volume per hour and to ensure that queues are managed optimally.

Figure 5.11 illustrates the different calculations and models that were used for determining the optimal number of voting terminals.

6.5.2 Vote Anywhere

Vote Anywhere was a feature adopted in the UAE FNC e-voting system to facilitate voting at any polling station without geographic limitations. Technically, a UAE voter could vote in any polling station across the country. However, as per UAE regulations, voters could vote only in their respective Emirates. Thus, the Vote Anywhere feature enabled voters to vote in any of the polling stations in their respective Emirates. For example, voters in Abu Dhabi had the option to vote in any of the four polling stations. The use of the UAE national identity card provided higher trust and confidence to enable this feature. The UAE national identity card was the primary identification document that provided authenticated access to the voting system. Once a vote was cast, the ID card chip was updated with a flag of 'voted' and it could not be used for voting again. The ID verification system was synchronised across the polling stations in each Emirate with information on voter verification. The verification system was available in real time in every polling station. The verification terminals were the starting point for voting, and only when cleared were voters able to proceed to the voting terminal.

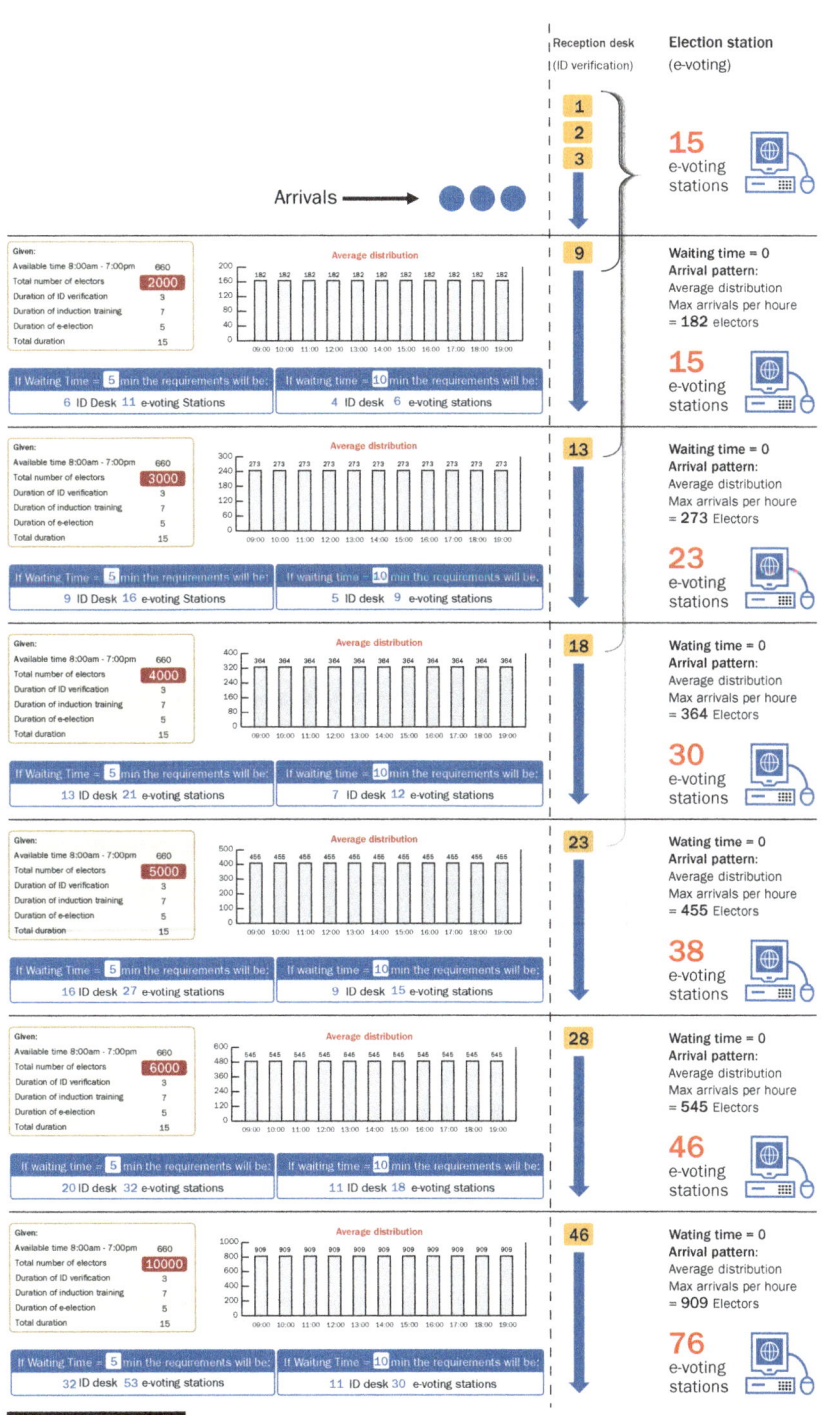

Figure 5.11   Calculation scenarios

6.5.3 Polling station layout

NEC determined an effective layout for each polling station to facilitate smooth flow of traffic during the election day (see Figure 5.12). This ensured that only the relevant people had access to relevant areas in the polling station. This was also supported by a robust technical design

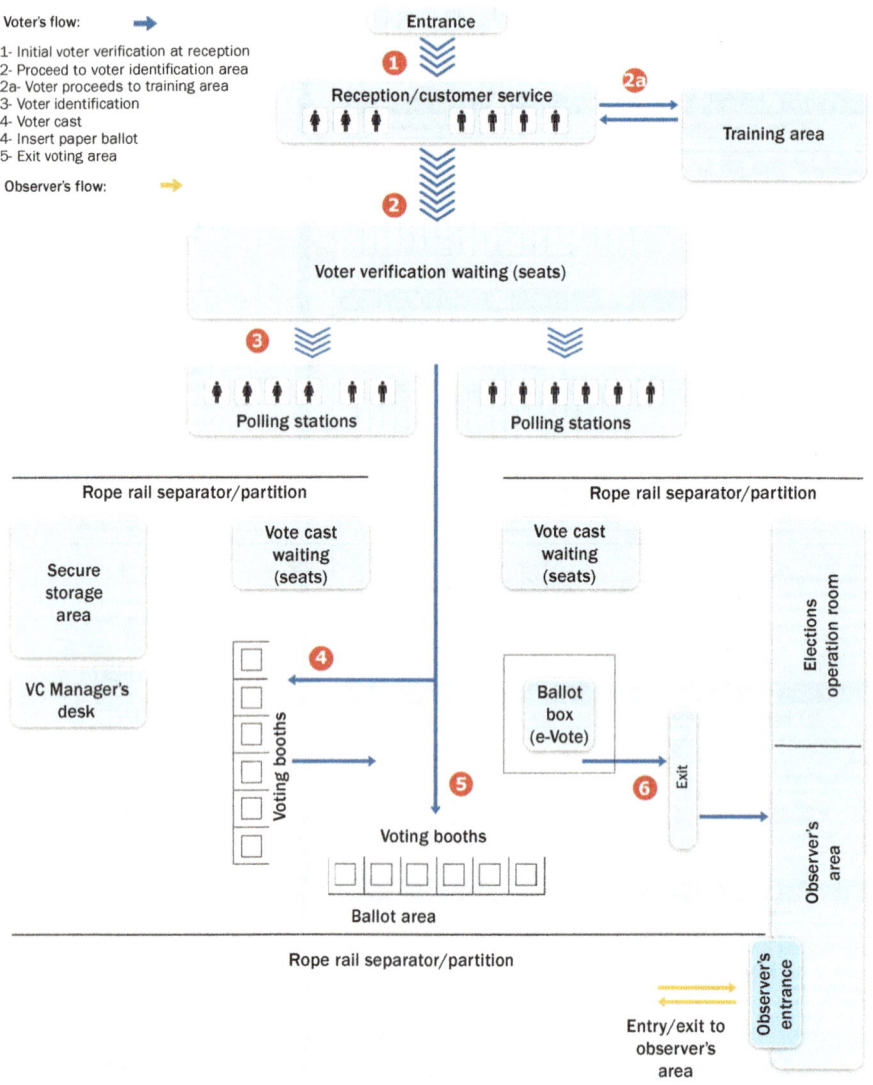

Figure 5.12   Polling station layout

set up to ensure accessibility and availability on election day. Figure 5.13 depicts one of these network designs.

On election day, more than 400 volunteer staff were stationed at the polling stations with various roles to support the electronic voting process. Some of these staff were responsible for training voters on the electronic voting system.

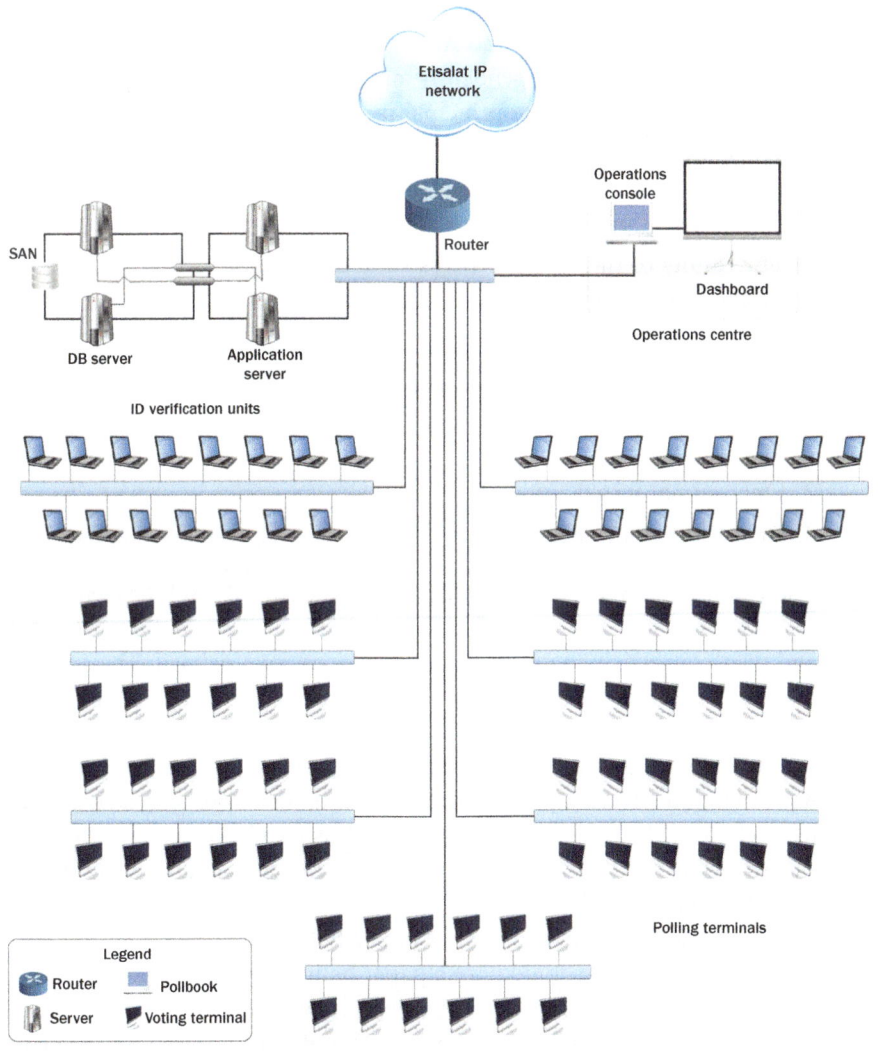

Figure 5.13 Voting site network design

6.6 Legal requirements

The e-voting system needed to meet strict security and privacy requirements, and comply with specific constitutional, legal and regulatory contexts related to the electoral rules in the UAE. In general, the following electoral rules were defined:

- Each polling station to have an Electoral Chief Officer responsible for the polling station.
- An Electoral Board (EB) and Administration Board will be established for each Emirate independently and the polling stations' Electoral Chief Officer will be the principal member.
- The counting process should be initiated and performed in the presence of the candidates of the corresponding Emirate for transparency purposes.
- The results of the voting process should be announced and published by the Electoral Chief Officer of each Emirate.

In addition, three legal systems elements were identified and dealt with in detail for the elections. These elements were:

1. Voting systems.
2. Operational instructions.
3. Department of appeals.

The voting systems detailed the type of voting terminals and the technology requirements of the overall voting system. The polling stations and their connectivity with the central locations were defined. The e-voting infrastructure was laid out. Operational Instructions detailed the voting process. 'one voter one vote' was the defined voting policy. The number of representative seats for each Emirate was defined using proportional representation. Thus, Abu Dhabi and Dubai had four candidates each to be elected. The Emirates of Sharjah and Ras Al Khaimah had three to be elected, while the Emirates of Ajman, Umm Al Quwain and Fujairah had two each to be elected.

It was also mandated to have a paper ballot print-out with the vote cast so that the voting could have a paper ballot trail for count validation. The Department of Appeals allowed appeals for any disputes, or challenges in the post-election phase. It was of paramount importance to ensure voter secrecy. It was thus mandated that the voting system should separate the voter's information from the vote and that there should be no way of linking the two.

6.7 Switching from e-voting to paper votes

A contingency procedure was documented and put in place to address any requirement to switch from e-voting to manual paper voting. The procedure was required to be authorised by electoral authorities. In this case, the steps to follow were:

- The voting centre officer should announce the situation to the candidates, as well as the voters and the public witnessing the event.
- The polling station officer should also announce that the Vote Anywhere option is disabled and only a voter assigned to that polling station will be authorised to cast a vote.*
- Assistant technicians will verify that e-votes cast were synchronised with the central server.
- The polling station officer will authorise the procedure and technicians will shut down the e-voting solution.
- Poll-workers receive the pre-printed clean ballot papers to be handed to the voters.
- Poll-workers also receive the authorised list of voters for that polling station (from the poll-book application) including the information of the voters who have already cast their vote.
- Poll-workers will resume the voting process, providing the voter with a paper ballot instead of the ID required to vote (see Figure 5.14).

6.8 Security features in the system

The analysis conducted by the technical committee concluded that the overall security of the e-voting system is dependent on the level of protection provided by the operational processes put in place, rather than the security features of the system itself. This is also in compliance

---

* When authorised, the 'Vote Anywhere' system was to be procedurally disabled owing to the non-availability of the e-voting system. Keeping this contingency in mind, voters were mapped to the voting stations nearest to their homes. Thus, for example, all Al Ain-based voters were pre-mapped to the polling station in Al Ain, while voters in Jumeira city were mapped to the polling station in the Dubai World Trade Centre. Poll workers were trained to refer to the voter list, and direct the voters to their respective polling stations.

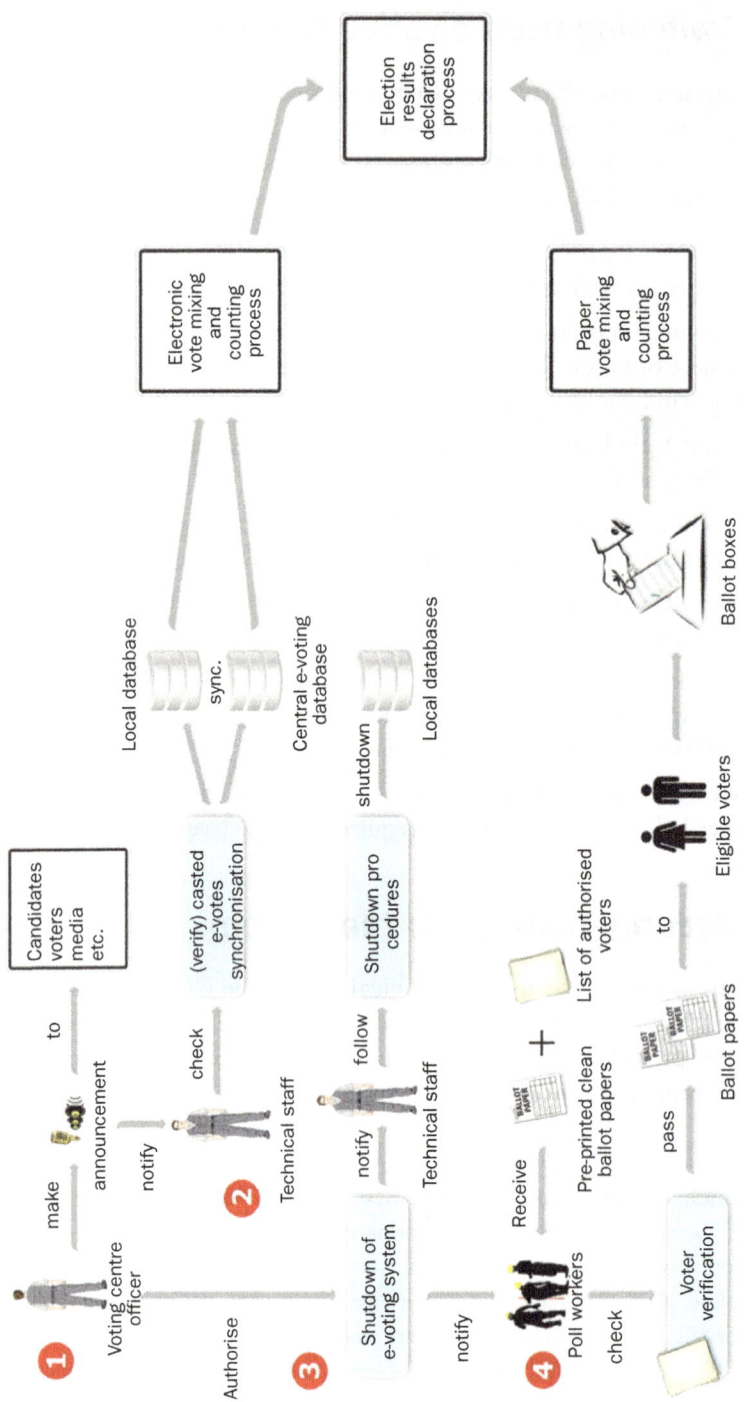

Figure 5.14 Procedure for switching to paper voting

with the findings of Hisamitsu and Takeda (2007), who stated that the security issues in e-voting systems arise from lack of protection mechanisms and procedures on tabulation machines. Each and every one of the almost 800 polling stations and 300 workers' laptops were limited, i.e. the basic software was restricted and their connectivity features limited so they could only be used for the election.

Overall, the computerised e-voting system was required to meet all the security aspects of manual voting. The committee defined the following measures to be implemented to support the overall security of the system.

6.8.1 Software certification

- The e-voting software itself was an international commercial e-voting solution that was based on proprietary source codes. The e-voting system was required to be certified by an international body. The core e-Voting Solution was certified by an independent international body following EU guidelines, namely the Austrian Centre for Secure Information Technology (A-Sit), an independent renowned certification authority appointed by the Federal Ministry of Science and Research to certify the security of the software. An audit was carried out applying the e-voting standards of the European Council as well as analysing the security architecture and source code of the e-voting software in accordance with Austrian law. The used software development processes and methodologies were audited against ISO/IEC 15408. Certification process guaranteed that the machines and software were reliable and secure.

6.8.2 Security of polling station equipment and communication

- Network segregation was used for the local networks at the voting sites, including networks for voting terminals and poll-books, database networks, and external network connections.
- All servers use local firewalls.
- All operational machines (laptops and voting terminals) were locked to allow only authorised changes to the software installed on these machines.

6.8.3 Data security

- SSL connections were established between voting terminals and application servers.
- Voting options were encrypted at the voting terminal with a digital envelope using the Election Public Key.
- The digital envelopes containing the voting options were digitally signed using anonymous digital certificates.

6.8.4 Voter authentication

- Voters were authenticated using their biometrics stored in national electronic ID cards.
- A secure flag was recorded on the ID card to indicate that the voter had been verified and was then allowed to vote at the same site.
- The verified card could then be used to vote.
- At the voting station, a digital certificate was used to authenticate the voter and to secure the voting session.
- The digital certificate was also used to encrypt the electronic vote.
- Once the vote was cast in the voting system, the card was flagged to prevent the voter from accessing the voting system again.

6.8.5 Voter anonymity

- The encrypted votes were digitally signed with anonymous digital certificates.
- The signed votes were decrypted only using the election private key, which was constructed from the individual keys of the electoral board members. This step was done after the voting was complete in order to start the count.
- During mixing, the system broke any correlation between the encryption envelope and the vote content. The outcome of this step was that the contents of all votes were cast into the system during voting, with no links whatsoever between each vote and the identity of the voter.
- Temporal or residual information managed by the voting applications (e.g. cookies or temporal records) were destroyed after the vote was cast, removing any possible trace containing voter information or vote selections.

6.9 Pilot test

The e-voting system was piloted two weeks prior to election day. More than 600 volunteer staff from the EIDA participated in the testing of the system. Although no major technical issues were identified, many of the reported issues were related to usability and overall organisation. Much attention was needed to be given to the training of support staff on election day.

Meticulous planning went into pilot testing. All systems were set up at the actual voting site in Dubai. This was a mock system working in actual voting conditions. The pilot itself was a culmination of the factory acceptance tests, followed by detailed user acceptance tests (UAT) conducted on the e-voting systems. The test scripts used for the UAT set the tone for pilot testing. The pilot was therefore an actual dry run prior to the elections.

The dry run served two purposes. The first was to test the resilience of the e-voting system, while the second was to ensure that human resources were trained and acquainted with the voting systems for election day. The scope of the dry run was also to test:

1. The complete e-voting solution required to run the voting process on election day with the required IT Infrastructure; and

2. The actual voting process as prescribed for the UAE elections 2011.

Dubai was chosen as the location for the dry run. The identified voting centre at the Dubai World Trade Centre was set up with coordination and cooperation from the DWTC team.

Table 5.6 outlines some additional elements that took place during the pilot test.

6.10 Election day

Early preparation activities started a few days in advance with election system configuration and final setup for each Emirate. This involved updating the e-voting system in each Emirate with voter and candidate lists, configuring the election start and end date/time for elections in each of the seven Emirates and so on. The electoral and administration board in each Emirate was responsible for authorising the beginning and end of the election in their Emirate.

The election commenced on 24 September 2011 at 8 am. Election day activities started with a team briefing by the polling station manager, after which designated and trained staff started the procedure of booting up the voting terminals and initiated the process of voter identification applications (see Figure 5.15). Technical and support staff were present

Table 5.6   Pilot test constituents

Objective	Description
Environment	A mock election scenario was set up in the local servers at the voting centre.
Test Material	1. A file dump was prepared with electoral roll and the candidates. 2. A mock election was defined and configured for a. Institution, Election Event and Election identification. b. Dates, start and end times. c. Electoral rules (replica of actual election rules). Users in each user role were identified and teams constituted to complete end to end voting process.
Equipment	Complete voting centre setup comprising of all equipment (servers, network, switches, voting terminals, voter identification terminals, ID card readers, biometric readers) required at the polling station on election day were put in place. Dry run participants used both their national ID cards and the contingent white smart cards to cast the votes.
Resources	The Emirates Identity Authority mobilised nearly 600 of its staff to participate in the dry run in addition to the people deployed from Takatuf (an organisation providing volunteers). A formal mock Electoral Board was constituted and the pilot election was opened for voting.
Communication	A key element in the pilot planning was communication. A detailed communication plan was worked out and resources were found to make the dry run a success.

Figure 5.15   Election day snapshot

at the polling stations to assist staff in electronic voting related matters.

Below is primary set of rules governing voting procedures related to what happened at polling stations (see Figure 5.16). Voters arriving at the voting centre were:

1. Guided by receptionists to training areas. Separate group and individual training areas were set up to familiarise voters and prepare them for the voting process.

2. Voters were validated at ID verification desks. Individuals arriving to vote were identified using their ID cards and validated against the electoral roll. Only valid voters were allowed to proceed to the voting area.

3. Valid voters cast their votes electronically at the voting terminals. Voters used their ID cards or smart cards prepared for voting to cast their electronic vote. The system only allowed a voter to vote once. Once the electronic vote was cast, a printed copy of the ballot was printed.

4. Voters were instructed to fold and insert the ballot copy in one of the ballot boxes and then exit the voting centre.

As mentioned previously, each polling station had its own local servers to maintain the record of votes cast during the day. For those Emirates that had more than one polling station, ballots were synchronised within the same Emirate to maintain a view of participation. Aggregated participation information from each Emirate was sent to the main Abu Dhabi site periodically to display aggregated participation information. The whole voting process throughout the country was monitored by the higher commission members at the operations management centre, which was set up to handle the technical and organisational issues that arose.

Ballot counting started when the election ended at 8:00 pm and the electoral board of the Emirate authorised the process. An additional hour was granted to Abu Dhabi due to growing numbers of voters in the last few hours. The results of each Emirate were announced in the main polling station of the Emirate. Overall, more than 35 thousand citizens participated in casting their ballots electronically. Figure 5.17 depicts the survey results conducted by some international research companies that outline the overall perception of citizens around FNC elections.

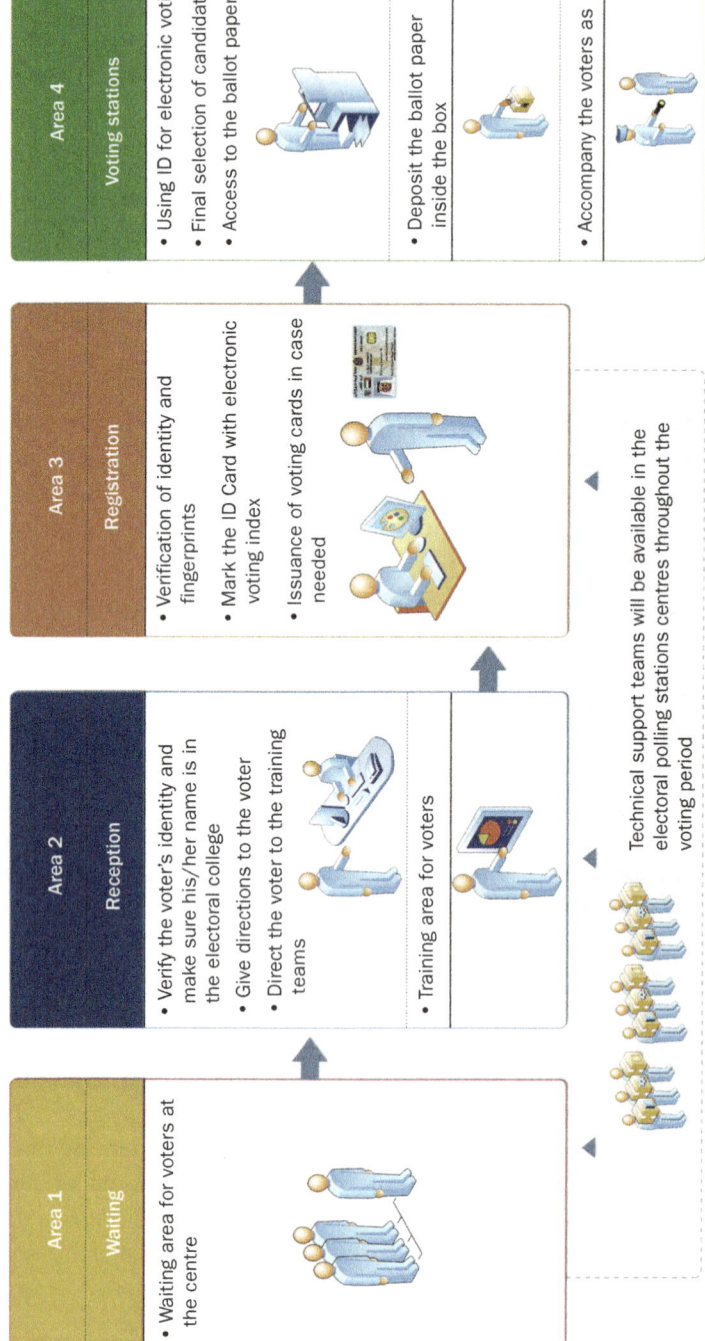

Area 1	Area 2	Area 3	Area 4
Waiting	Reception	Registration	Voting stations

Area 1 — Waiting
- Waiting area for voters at the centre

Area 2 — Reception
- Verify the voter's identity and make sure his/her name is in the electoral college
- Give directions to the voter
- Direct the voter to the training teams
- Training area for voters

Area 3 — Registration
- Verification of identity and fingerprints
- Mark the ID Card with electronic voting index
- Issuance of voting cards in case needed

Area 4 — Voting stations
- Using ID for electronic voting
- Final selection of candidates
- Access to the ballot paper
- Deposit the ballot paper inside the box
- Accompany the voters as

Technical support teams will be available in the electoral polling stations centres throughout the voting period

Figure 5.16 Voting procedures during election day

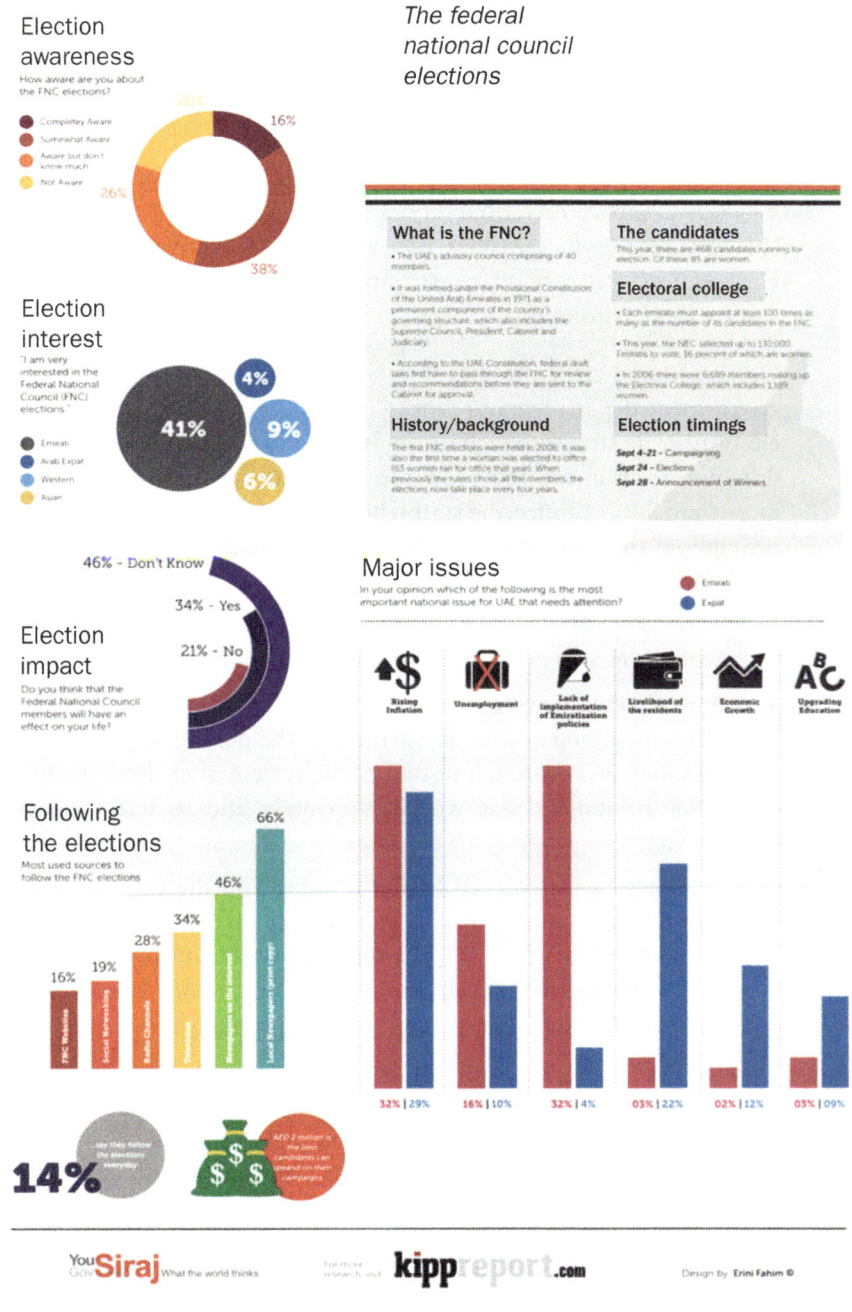

Figure 5.17 Kippreport survey results

7. Success factors

From a technical perspective, the use of advanced cryptographic protocols enabled all types of election processes to be carried out in a completely secure, transparent and auditable manner. The solution accorded the highest levels of security in terms of voter 'privacy, ballot box integrity, and voter verifiability'.

The key success factors that have led to the success of the FNC e-voting experience were related to 'structure and readiness', which can be summarised as follows:

1. Organisational structure played a key role in getting together many government departments to work cooperatively to support the FNC elections. The structure included more than 100 officials representing different government and public sector organisations, in addition to more than 900 volunteering staff who supported the 14 voting centres throughout the country.

2. Site readiness: a comprehensive detailed plan (Go-live plan) was defined to govern the prerequisite activities for the elections day. Great attention was given to election site preparation.

3. System readiness: the go-live plan mentioned above included activities related to the voting system deployment in the polling stations. These activities included thorough testing of the system after deployment to ensure all equipment was working properly and to test the sites' connectivity.

4. Staff readiness was achieved by the training and the dry run (pilot) simulation. A comprehensive training plan was defined and implemented to cover these activities. Staff training in the new methods of voting to a level good enough to provide on-site voter education and process knowledge-gathering provided valuable input to the election day.

5. Communication was recognised to be the most critical success factor. The key to strong communication was the ability to define the process and timescales at the outset. There was constant communication at every level of the election planning and execution process with the stakeholders through public meetings, workshops, local conferences and so on.

6. The use of social media and social networks supported the success of the FNC elections, as it turned communication into an interactive

dialogue between the NEC commission, local communities and individuals. Social media played an important role in widening accessibility and scalability of communications.

In addition to the above critical success factors, the application of the UAE population register and smart identity card was considered to have added significant contribution to the overall success of the NFC elections.

7.1 Using the UAE Population Register in the voting system

1. The UAE National Register was used to extract the list of the electoral colleges. The Register is the most accurate and unique population database in the country, as it depends on advanced technologies to link biographical information to individuals, namely facial and fingerprint biometrics

2. The voter list was uploaded to the voting system before the elections. This list was used to verify voters and ensure only legitimate voters were using the voting terminals to cast their electronic votes.

3. In addition to the voter list, the National Register was also used to generate the list of candidates, which was also uploaded to the voting system.

4. General voter information available on the national register was also used in the voter list. The information included age, gender, profession, Emirate and so on. This information was used by the voting system to provide statistical reports and dashboards during and after the elections.

5. The mobile numbers available in the national register were also useful to send reminder SMS messages to the voters before and on election day.

7.2 Using the ID card for voting

The UAE biometric-based national identity card was an important component which made secure e-voting possible. The importance of the electronic ID card is summarised by the following points:

1. The card was used to verify the identity of voters. Biometric verification was used in the verification process, and in cases when physical verification using picture comparison was not possible.

2. The card was also used to ensure that only verified voters could access the voting terminals.

3. After the vote was cast in the system, the card and e-voting system were flagged to ensure the voter could not use his card to vote again.

4. In the above process, vote buying was not possible, as was the case in other voting systems where electronic ID cards were not used. This was due to the fact that only the card owner could use his/her card to vote after being verified.

8. Lessons learned

The UAE e-voting system was conceived, designed and deployed, and worked exceptionally well albeit with some glitches and operational issues. Although it did not pose any serious risk to the overall and final election results, a few issues need to be taken note by election officials during planning and execution phases.

Some of the major challenges faced on election day were:

1. Some of the electronic voting systems developed unanticipated screen freezes, leaving voters wondering whether their ballots had been properly recorded. Despite this situation, voters were able to try to cast their ballot electronically again, but they got error messages informing them that they had already voted or they were not allowed to vote again. On-site support was provided to check whether the ballot was cast correctly or not.

2. Some printers went into sleep mode, which needed supervisor intervention to print the votes cast. Some printers were on default settings to go into sleep mode if unattended for a few minutes. The technical staff needed to visit each and every printer to change the default settings. With these printers in sleep mode, the ballots were not successfully printed but they were stored electronically. A reprinting feature was accessible with a specific password, although it seemed to be unknown by some participants. This feature is further described in Appendix 2.

3. Some voters by-passed the voter identification and went straight to the voting terminals only to find that the system did not allow them to cast their votes, leading to confusion among the voters.

4. During the mixing and counting, data synchronisation between two major polling stations stalled and it took some time for the data

synchronisation to be completed. This delayed the announcement of the election results in the Emirate of Abu Dhabi for about an hour. The issue was related to the MPLS data link between Abu Dhabi and Al Ain that had some problems and the ballot synchronisation speed was slow. This affected the mixing process, as it could not be executed until all the information was synchronised. This was not a mixing problem, but a data-readiness problem i.e., the pre-requisite of the mixing process was to have all the information synchronised.

5. Some voters' identity could not be verified when they used their national ID cards; thus, they were issued with white smart cards at polling stations to enable them to cast their votes. This was defined as a fallback scenario for voters whose identity card was damaged and their chip was not readable. This scenario is further described in Appendix 2. Thanks to having this scenario defined and implemented, all voters were able to cast their ballots. A similar verification process was executed with these voters.

Serious as they seem, all these challenges were overcome quickly due to the contingency measures put in place in the e-voting system and the overall procedures. For every issue reported, trained technical staff identified the root cause and set in motion contingency plans to enable the restoration of the voting process.

The local area network was quickly brought up and the voting system was restored for casting votes where screen freezes were reported. The screen freeze was attributed to the inability of the LAN to load the voting applet onto the voting terminals from the server.

Specifically during the first three hours, management and technical staff were on the move in polling stations to ensure that the printers were kept active throughout the voting period. Stricter controls were enforced ensuring that the voters did not approach the voting terminals without their identity being verified, thus enabling a smooth voting process. The network was fine-tuned to ensure data synchronisation between the affected sites.

As indicated above, and where the national ID card could not be verified, authorised voters were provided with 'white cards' enabling them to cast their votes on the e-voting terminals. Interestingly, the majority of those people who had shown up on election day and reported that their national ID cards did not work, were later recognised as not being on the electoral roll in the first place. As a result, those individuals needed to be dealt with very carefully.

Another important lesson learned related to the fact that the UAE e-voting system only printed voters' copies that were put in the ballot boxes by the voters. No separate receipts were printed for the voters' own records. Some of the voters expressed concern and wondered if their votes were actually counted. It is therefore recommended that, for future elections, the NEC should put in place kiosk machines with an application to allow voters to check whether their votes were counted after the mixing stage. Instead of a separate receipt, national identity cards could be used as a token to be inserted into such kiosk machines for electronic verification. This step is envisaged to enhance voters' confidence and levels of trust.

Taken as a whole, these contingency measures, combined with on the ground leadership from top management (Higher Election Commission members), ensured a successful culmination of this ambitious project of e-voting in the UAE. The following stand out as the main highlights of the UAE FNC elections exercise:

- Successful completion of the elections using e-voting.

- All voting results for winners announced the same day and within hours of the closing of voting.

- Successful deployment of national ID card for ID verification.

- Tremendous coordination in organising, deploying and running the e-voting system, involving many entities and individuals across the country. This more than amply demonstrates the leadership qualities and the organising abilities of the project management team and the project sponsoring committee, i.e. the NEC.

The success of this e-voting system has laid the foundation for the UAE to step forward towards full i-voting system deployment. The UAE now has all the technological and technical components required to make this a reality. These are as follows:

1. A proven national ID card that carries the digital certificates from the population PKI that should enable identification, validation, authentication, digital signing and encryption.

2. A proven e-voting system that was designed based on browser based voting on PCs. The voting terminals were designed as kiosks with touch screen PCs embedded in the kiosks, with ID card readers and biometric units.

3. A proven e-voting system that has been successfully deployed and fully integrated with the national ID card as a verification tool.

4. A proven nation-wide interconnectivity and bandwidth that enables remote voting.

5. Robust e-voting practice and process that enables secure and reliable voting remotely.

6. A legal system that has had experience with using national ID cards as the primary token for identification.

9. Conclusion

As computing and security technologies have evolved, so too has the adoption of e-voting technologies for conducting national elections. Technologically advanced countries and countries with electoral maturity have set the pace for the adoption of e-voting. Although countries have had different experiences with electronic voting systems, there seems to be a consensus among governments on the importance and the positive impact of electronic voting systems on the overall election systems. The literature identified numerous advantages of electronic voting systems over traditional paper ballot voting, namely convenience, usability, simplicity, cost savings, reliability and so on. These factors explain to some extent the underlying reasons behind the growing interest of governments to deploy electronic voting systems.

In this article, we attempted to provide an overview of elections and e-voting systems. We covered the existing literature to review the critical points of current knowledge, including substantive findings and contributions to the particular topic of e-voting. The article then moved to provide an overview of the UAE FNC elections in detail. The presented research content is considered to be of great support to similar endeavours, specifically as countries worldwide are showing greater interest in the application of e-voting systems.

In general, the FNC 2011 elections were seen as one of the most successful initiatives of the year in the UAE. The e-voting system was considered by the government to be a major improvement to the manual voting process in terms of:

- Integrity: the e-voting system is an extremely secure system that cannot be tampered with by anyone, including system administrators.

- Speed: vote counting and the announcement of results were done based on the electronic votes in the system within less than an hour after closing the polling stations.

- Transparency: the entire voting process was observed by the candidates and their representatives as well as the media. The system was open for audit by external and internal auditors. Vote counting was done automatically and entirely by the system. Preliminary results from the system were presented in each of the 14 polling stations that promoted transparency, and allowed candidates to see them before the official announcement of results.

The voting procedures announced through different media channels played a major role to encourage participation and turnout. The use of the government-issued advanced smart national identity cards to identify the voters and candidates acted as an effective enabler of the e-voting system. By and large, the automated e-voting system enabled smooth and fast elections across the seven Emirates to be conducted and closed on the same day.

9.1 Future work

Research in the field of e-voting is an important factor in improving overall knowledge, where it can provide much valuable information from sharing the results of different pilots and experiments in countries worldwide. Such a range of experience provides a wealth of practical information, knowledge and input. The content of the research and the lessons learned from the deployment of e-voting in the FNC election in the UAE can serve as a set of valuable guidelines for the future design and deployment of e-voting systems in Arab countries and worldwide.

Indeed, the case study approach adopted in this research has its limitations; generalisability is not possible due to the information source being from a single case. However, it needs to be noted that the UAE e-voting system design and procedures from technical, legal, management and other aspects were benchmarked with their implementation in other countries. This reduces and addresses the concerns over case study limitations.

The government of the UAE is in the process of trialing an Internet-based e-voting system. This application will be available for deployment for any election requirement in the public sector. The system will rely greatly on the smart identity card's capabilities for strong authentication. The public key cryptography will form the foundation on which secure communications will be established. The government trusts that biometrics, smart cards, and national cryptography will, together, provide strong mechanisms to protect the integrity of voter registration information and overall election processes.

In our opinion, the use of government-issued smart cards adds a new dimension of security and contributes to the overall trust and confidence of the public. More research is inevitable, such as the Internet, in order to explore how such cards could support the development of more secure electronic voting over public networks.

References

1. Alexander, K. (2001) 'Ten Things I Want People to Know About Voting Technology'. Presented to the Democracy Online Project's National Task Force, National Press Club, Washington DC, 18 January 2001.
2. Schoenmakers, B. (1999) 'A simple publicly verifiable secret sharing scheme and its application to electronic voting'. *Advances in Cryptology-CRYPTO* 1666 Lecture Notes in Computer Science: 148–164. Berlin. Springer-Verlag.
3. Barr, C. W. (2006) 'Security Of Electronic Voting Is Condemned, Washington Post'. Available at: *http://www.washingtonpost.com/wpdyn/content/article/2006/11/30/AR2006113001637.html*
4. Bonsor, K. and Strickland, J. (2007) 'How E-voting Works: The Psychology of Electronic Voting'. *http://people.howstuffworks.com/e-voting5.htm*
5. Bonsor, K. and Strickland, J. (2011) 'How E-voting Works'. *http://www.howstuffworks.com/e-voting.htm*
6. Caltech-MIT (2001) 'A Preliminary Assessment of the Reliability of Existing Voting Equipment'. The Caltech-MIT Voting Technology Project, March 30, 2001. Available at: *http://www.vote.caltech.edu/Reports/index.html*
7. Chaum, D. (2004) 'Secret-Ballot Receipts: True Voter-Verifiable Elections'. *IEEE Security and Privacy* 2 (1): 38–47.
8. Chiang, L. (2009) 'Trust and security in the e-voting system'. *International Journal of Electronic Government* 6 (4): 343–360.
9. CIVTF (2000) 'A Report on the Feasibility of Internet Voting'. California Internet Voting Task Force. Available at: *http://www.sos.ca.gov/elections/ivote/final_report.pdf*
10. Clausen, D., Puryear, D. and Rodriguez, A. (2000) 'Secure voting using disconnected distributed polling devices'. Palo Alto. CA: Stanford University. Available at: *http://www-cs-students.stanford.edu/~dclausen/voting/cs444n_voting_report.pdf*
11. Crane, R., Keller, A., Dechert, A., Cherlin, E. and Mertz, D. (2005) 'A Deeper Look: Rebutting Shamos on e-Voting'. Available at: *http://www.verifiedvoting.org/downloa-ds/shamos-rebuttal.pdf*
12. Cranor, L. F. (2000) 'Voting After Florida: No Easy Answers'. Available at: *http://lorrie.cranor.org/voting/essay.html*
13. Done, R. S. (2002) 'Internet Voting: Bringing Elections to the Desktop'. Research Report, Pricewater house Cooper's Endowment for the Business of Government. Available at: *http://www.endowment.pwcglobal.com/pdfs/Done_Report.pdf*

14. e-Gov Monitor (2003) 'Does the UK need e-voting?' Available at: *http://www.egovmonitor.com/features/evoting2003.html*

15. Britannica (2012). Encyclopedia Britannica. Available at: *http://www.britannica.com*

16. E-Poll (2012). Available at: *http://www.e-poll-project.net*

17. González, J. F. and Brambila, S. B. G. (2012) 'Secure Architectures for a Three-Stage Polling Place Electronic Voting System'. *Computación y Sistemas* 16 (1): 43–52.

18. Hargrove, T. (2004) 'Widespread voting woes foil democratic process. Old equipment, faulty accounting methods fail to tabulate many votes'. The Detroit News.

19. Hisamitsu, H. and Takeda, K. (2007) 'The security analysis of e-voting in Japan'. Alkassar, A. and Volkamer, M. (eds), 'e-Voting and Identity'. *Proceedings of VOTE-ID'07 1st international conference on E-voting and identity*: 99–110. Berlin, Heidelberg, New York: Springer-Verlag.

20. Hoffman, L. J. (2004) 'Internet Voting: Will it Spur or Corrupt Democracy'? Technical Report, Computer Science Department, The George Washington University, Washington, DC. Available at: *http://www.cfp2000.org/papers/hoffman2.pdf*

21. Hout, M., Mangles, L., Carlson, J. and Best, R. (2004) 'The Effect of Electronic Voting Machines on Change in Support for Bush in the 2004 Florida Elections'. Available at: *http://ucdata.berkeley.edu/new_web/VOTE2004/election04_WPwappendices.pdf*

22. IDEA (2011) 'Introducing Electronic Voting: Essential Considerations'. Policy Paper, The International Institute for Democracy and Electoral Assistance. Available at: *http://www.agora-parl.org/sites/default/files/e-voting_idea.pdf*

23. Ikonomopoulos, S., Lambrinoudakis, C., Gritzalis, D., Kokolakis, S. and Vassiliou, K. (2002) 'Functional Requirements for a Secure Electronic Voting System'. *Proceedings of the 17th IFIP International Conference on Information Security*: 507–520. Egypt: Kluwer Academic Publishers. Available at: *http://www.instore.gr/evote/evote_end/htm/3pub-lic/doc3/public/aegean/paper4.pdf*

24. IPI (2001) 'Report of the National Workshop on Internet Voting: Issues and Research Agenda'. Internet Policy Institute. Available at: *http://verifiedvoting.org/downloads/NSFInternetVotingReport.pdf*

25. Jones, D. W. (2001) 'Evaluating Voting Technology'. Testimony before the United States Civil Rights Commission, Tallahassee, Florida. Available at: *http://homepage.cs.uiowa.edu/~jones/voting/ArizonaDist20.pdf*

26. Kaplan, A. M. and Haenlein, M. (2010) 'Users of the world, unite! The challenges and opportunities of Social Media'. *Business Horizons* 53 (1): 59–68.

27. Klein, P. (1995) 'An Untraceable, Universally Verifiable Voting Scheme'. Seminar in Cryptology, December 12, 1995.

28. Kosmopoulos, A. (2004) 'Aspects of Regulatory and Legal Implications on e-Voting'. *Lecture Notes in Computer Science* 3289: 589–600.

29. Law, G. (2002) 'Britain back on e-voting track. UK city council begins smartcard e-government plan'. *PC Advisor* 25 April.

30. Levy, S. (2004) 'Ballot Boxes Go High Tech'. *MSNBC Newsweek*.
31. Liptrott, M. (2006) 'e-Voting in the UK: a Work in Progress'. *The Electronic Journal of e-Government* 4 (2): 71–78.
32. Manjoo, F. (2002) 'Voting into the void New'. *Salon* 5 November. Available at: *http://www.salon.com/2002/11/05/voting_machines_2/*
33. Manjoo, F. (2003a) 'Another case of electronic vote-tampering'. *Salon* 29 September. Available at: *http://www.salon.com/tech/feature/2003/09/29/voting_machine_standards*
34. Manjoo, F. (2003b) 'An open invitation to election fraud'. *Salon* 23 September. Available at: *http://www.salon.com/tech/feature/2003/09/23/bev_harris*
35. Mercuri, R. (1993) 'The Business of Elections'. *CFP'93*. Available at: *http://www.cpsr.org/conferences/ cfp93/mercuri.html*
36. Mercuri, R. (2002) 'A Better Ballot Box'? *IEEE Spectrum* 39 (10): 46–50.
37. Millar, S. (2002) 'Don't trust computers with e-votes, warns expert'. *The Guardian* October 17. Available at: *http://www.guardian.co.uk/politics/2002/oct/17/uk.internet*
38. Neumann, P. G. (1993) 'Security Criteria for Electronic Voting'. 16th National Computer Security Conference, Baltimore, Maryland, September 20–23.
39. NIST (2006) 'Requiring Software Independence in VVSG 2007: STS Recommendations for the TGDC'. The National Institute of Standards and Technology (NIST). Available at: *http://vote.nist.gov/DraftWhitePaper OnSIinVVSG2007-20061120.pdf*
40. NIST (2008) 'A Threat Analysis on UOCAVA Voting Systems'. *NISTIR* 7551, The National Institute of Standards and Technology (NIST). Available at: *http://www.nist.gov/itl/vote/upload/uocava-threatanalysis-final.pdf*
41. NIST (2011) 'Security Considerations for Remote Electronic UOCAVA Voting'. *NISTIR* 7770, The National Institute of Standards and Technology (NIST). Available at: *http://www.nist.gov/itl/vote/upload/NISTIR-7700-feb2011.pdf*
42. O'Donnell, P. (2004) 'Broken Machine Politics'. *Wired Magazine*, January 2004. Available at: *http://www.wired.com/wired/archive/12.01/evote_pr.html*
43. Oostveen, A. and P. van den Besselaar (2004) 'Security as Belief. User's Perceptions on the Security of Electronic Voting Systems'. In: Prosser, A. and Krimmer, R. (eds). *Electronic Voting in Europe: Technology, Law, Politics and Society*. Lecture Notes in Informatics 47: 73–82. Bonn: Gesellschaftfür Informatik. Available at: *http://www.social-informatics.net/ESF2004.pdf*
44. Oostveen, A. and van den Besselaar, P. (2004) 'Ask No Questions and Be Told No Lies. Security of computer-based voting systems: trust and perceptions'. In Gattiker, U. E. (ed). EICAR 2004 Conference. Copenhagen: EICAR e-V.
45. Oostveen, A. and van den Besselaar, P. (2004) 'Internet voting technologies and civic participation, the users perspective'. *Javnost/The Public* 11 (1): 61–78. ISSN 1318–3222.
46. Oostveen, A. and van den Besselaar, P. (2004) 'E-democracy, Trust and Social Identity: Experiments with E-voting technologies'.

47. Oostveen, A. and van den Besselaar (2005) 'The Effects of Voting Technologies on Voting Behaviour: Issues of Trust and Social Identity'. *Social Science Computer Review* 23 (3): 304–311. Sage Publications.

48. Oostveen, A. and van den Besselaar, P. (2006) 'Non-Technical Risks of Remote Electronic Voting'. Ari-Veikko, A. and Mattia, M. (eds). *The Encyclopedia of Digital Government*: 502–507. Idea Group Inc.

49. Phillips, D. and von Spakovsky, H. (2001) 'Gauging the risks of Internet elections'. *Communications of the ACM* 44 (1): 73–85.

50. Pratchett, L. and Wingfield, M. (2004) 'Electronic voting in the United Kingdom. Lessons and limitations from the UK Experience'. In Kersting, N. and Baldersheim, H. (eds). *Electronic Voting and Democracy. A Comparative Analysis*. London: Palgrave Macmillan.

51. Prosser A. and Krimmer, R. (eds). (2004) 'Electronic Voting in Europe: Technology, Law, Politics and Society'. *Lecture Notes in Informatics* 47: 73–82. Bonn: GesellschaftfürInformatik.

52. Quesenbery, W. (2003) 'Starting from People: Usability and User-Centred Design in Voting Systems'. NIST Symposium on Building Trust and Confidence in Voting Systems. Available at: *http://www.nist.gov/itl/vote/upload/6-Quesenbery.pdf*

53. Cramer, R., Franklin, M., Schoenmakers, B. and Yung, M. (1996) 'Multi-authority secret ballot elections with linear work'. *Advances in Cryptology-EUROCRYPT'96, Lecture Notes in Computer Science* 1070: 72–83. Berlin: Springer-Verlag.

54. Rapoport, R. N. (1970) 'Three Dilemmas in Action Research'. *Human Relations* 23: 499–513.

55. Rivest, R. L. (2004) 'Electronic voting, technical report'. Laboratory for Computer Science, Massachusetts Institute of Technology. Available at: *http://theory.lcs.mit.edu/~rivest/Rivest-ElectronicVoting.pdf*

56. Rubin, A. D. (2002) 'Security Considerations for Remote Electronic Voting'. *Communications of the ACM* 45 (12): 39–44.

57. Salem, F. (2007) 'Enhancing Trust in e-Voting through Knowledge Management: The Case of the UAE'. Dubai School of Government. United Nations Public Administration Network. Available at: *http://unpan1.un.org/intradoc/groups/public/documents/unpan/unpan026090.pdf*

58. Saltman, R. G. (1975) 'Effective use of computing technology in vote-tallying'. The National Institute of Standards and Technology (NIST). Available at: *http://csrc.nist.gov/publications/nistpubs/NBS_SP_500-30.pdf*

59. Schneier, B. (2004) 'What's wrong with electronic voting machines'? *Open Democracy*. Available at: *http://www.opendemocracy.net/media-voting/article_2213.jsp*

60. Wolchok, S., Wustrow, E. Isabel, C. and Halderman, J. A. (2012) 'Attacking the Washington, D. C. Internet Voting System'. *Proceedings of the 16th Conference on Financial Cryptography & Data Security*. Available at: *https://jhalderm.com/pub/papers/dcvoting-fc12.pdf*

61. Shamos, M. I. (1993) 'Electronic Voting – Evaluating the Threat'. International Conference on Computers, Freedom, and Privacy, Burlingame, California.

62. Susman, G. L. and Evered, R. D. (1978) 'An Assessment of the Scientific Merits of Action Research'. *Administrative Sciences Quarterly* 23 (4): 582–603.
63. UAE NEC (2011) UAE National Election Commission. Available at: *http:// www.uaenec.ae*
64. UK POST (2001) 'Online Voting'. Postnote – a publication of the UK Parliamentary Office of Science and Technology.
65. Van den Besselaar, P. and Oostveen, A. (2003) 'E-voting is not neutral'! *Lecture Notes in Informatics* 35: 218–221.
66. Van den Besselaar, P. and Oostveen, A. (2004) 'Media effects in voting and polling: e-democracy, trust and social identity'. *Proceedings of the 2004 ICA Conference Communication in the Public Interest*, New Orleans. Available at: *http://citation.allacademic.com/meta/p_mla_apa_research_citation/1/1/3/0/1/ pages113010/p113010-1.php*
67. Volkamer, M. (2009) 'Evaluation of Electronic Voting'. Berlin: Springer-Verlag.
68. Waddington, D. (1994) 'Participant Observation'. In Cassell, C. and Symon, G. (eds). *Qualitative Methods in Organisational Research – A Practical Guide*. London: Sage.
69. Walsham, G. (1993) 'Interpreting Information Systems in Organisations'. Chichester: John Wiley & Sons.
70. Windley, P. J. (2005) 'e-Voting, Extreme Democracy'. In Lebkowsky, J. and Ratcliffe, M. (eds). *Extreme Democracy. Internet/Media Strategies, Inc*: 132–138. Available at: *http://www.extremedemocracy.com/chapters/ Chapter%2011-Windley.pdf*
71. Xenakis, A. and Macintosh A. (2006) 'A generic re-engineering methodology for the organized redesign of the electoral process to an e-electoral process'. In: Krimmer, R. (ed). *Electronic Voting 2006*. Lecture Notes in Informatics, 2nd International Workshop Co-organized by Council of Europe, ESF TED, IFIP WG 8.5 and E-Voting. CC. August, 2nd–4th in Castle Hofen, Bregenz, Austria. Available at: *http://www.e-voting.cc/wp-content/uploads/ Proceedings%202006/5.1.Xenakis_Macintosh_BPR_in_E-Voting_119-130. pdf*
72. Zetter, K. (2003) 'E-Vote Firms Seek Voter Approval'. *Wired News*. Available at: *http://www.wired.com/ news/evote/0,2645,60864,00.html*

Appendix 1

Functions and responsibilities of the National Election Commission and its sub-committees.

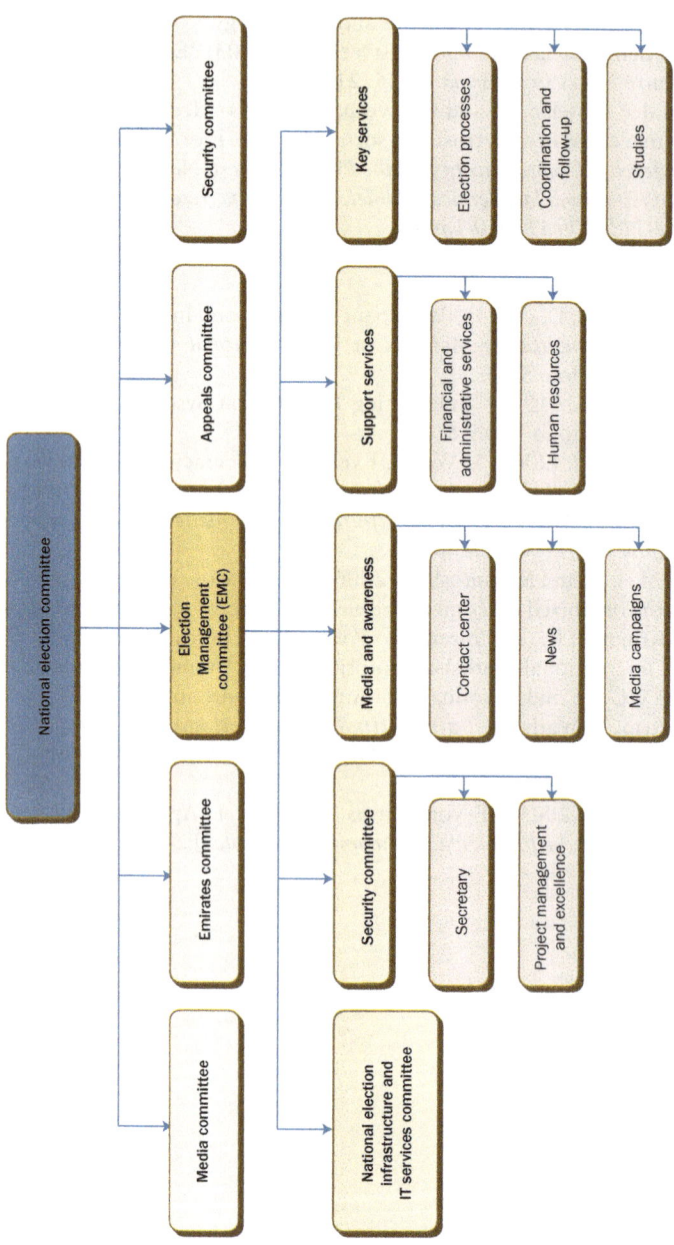

1. The National Election Commission (NEC)

The National Election Commission (NEC) was formed in February 2011 by a presidential decree consisting of government officials and public figures. It was empowered to oversee the whole election process, including:

- Setting out the overall election framework.
- Supervising the elections.
- Supporting efforts to raise electoral awareness.
- Writing down election guidelines.
- Locating polling stations in each Emirate.
- Approving regulatory measures for establishing the electoral legal framework.
- Issuing and seeking approval for the governing rules for the lists of Electoral Panel Members.
- Setting the date of elections.

2. The Election Management Committee

- Directing, supervising, and monitoring its staff as per applicable regulations and NEC decisions.
- Coordinating with relevant bodies, and coordinating subcommittee activities in these bodies, to ensure full implementation of prescribed duties and functions.
- Identifying manpower and financial needs required for implementing assigned tasks and reporting them to the NEC.
- Recommending necessary decisions and regulations for the functions of the Election Management Committee (EMC), and its subcommittees; and reporting them to the NEC for approval, and monitoring their implementation.

- Monitoring the implementation of NEC electoral decisions and guidelines directed to subcommittees concerning the preparation for elections, and reporting the same to the NEC in a timely manner.
- Preparing election budgets.
- Taking all necessary action to ensure the safety of elections.
- Assessing appeals for later submittal to the NEC.
- Submitting statements of votes to the NEC, before announcing final results.
- Any other functions assigned by the NEC.
- The Committee may take all necessary steps for the implementation of its assigned tasks. It may also procure the services of experienced professionals, as appropriate.

2.1 The Project Management and Excellence Committee

- Suggesting strategic and operational plans; and setting out and prioritising policies, and how to measure them.
- Preparing and developing these plans, along with work programs, in coordination with other relevant organisational units in the Committee.
- Designing performance indicators and submitting performance reports to all these units in the Committee and concerned parties.
- Monitoring strategic plans, evaluating performance and enhancing services at the EMC. The Office should also provide appropriate procedures, evidences, and systematic documentation for all operations, based on requirements.
- Managing projects based on established methodologies.
- Ensuring criteria are met in all measures taken.
- Handling the chairman of the Committee's reports and correspondence, whilst making all necessary arrangements and follow-ups for his meetings with relevant bodies.
- Measuring all concerned parties results and conducting compliance and research.

2.2 Infrastructure and IT Services Committee

- Implementing EMC-approved policies, strategies, standards, procedures, and regulations.
- Providing highly effective IT services to cope with election needs.
- Identifying IT program and system requirements to enhance networking connections as required for the Election Management Committee, whilst setting the IT work plan and needs.
- Designing and developing computer systems to be used for elections, and ensuring they meet preset standards.
- Setting the IT budget.
- Documenting the system throughout deployment and operation phases.
- Supervising the upgrading of IT networks required for making well-developed systems.
- Checking the specifications of networks and connection lines currently in operation.
- Preparing and presenting back-up plans in cases of emergencies.
- Training system users; informing them of their advantages and methods of operation.
- Technically supervising the website, in coordination with the Media and Communications and Electoral Processes Unit.
- Laying down different scenarios for e-voting, in terms of connection and IT requirements.
- Ensuring the readiness, safety, and security of IT systems.
- Providing all IT needs for media centres and polling stations.
- Any other functions assigned by the EMC.

2.3 Key services

2.3.1 Election processes

- Setting and implementing relevant policies, strategies, standards, and procedures for highly efficient electoral service provision.
- Preparing, saving, and classifying Electoral Panel members' records and databases.
- Informing Panel members of their electoral responsibilities.

- Registering Electoral Panel members.

- Updating Panel member databases.

- Organising and registering candidates.

- Identifying the appropriate method for announcing the Names of the Electoral Panel Members (Newspapers, phone calls, or SMS).

- Coordinating with the seven Emirates Rulers' courts to acquire the names of the Electoral Panel Members.

- Providing an appropriate form which includes all required data of the Panel Member. This form must meet electoral needs and requirements and facilitate research and studies.

- Launching and updating website pages displaying the names of the Electoral Panel Members in coordination with the Media and Communication Unit and the IT Committee.

- Furnishing the Contact Centre with the Electoral Panel Members' names.

2.3.2 Coordination and follow-up

- Streamlining communications and follow-up for different Emirates committees.

- Monitoring how far Committee decisions are put into action.

- Assessing and handling complaints and feedback, except for appeals.

- Working closely with election partners by showing good understanding and response to their needs.

- Providing everything necessary to enhance coordination between the Emirates committees and the Election Committee, whilst enacting relevant procedures by arranging Committee agendas, documenting their individual minutes, monitoring the implementation of recommendations, and fully coordinating with relevant bodies concerning the functions of the Emirates Committees Coordinator, and any other relevant coordinating functions.

2.3.3 Studies

- Setting goals and plans for research and studies according to Committee requirements, whilst supervising the implementation of each.

- Providing research, advisory, and consultation services concerning electoral affairs.

- Conducting research, studies, questionnaires, and evaluation forms that aim at measuring public opinion trends concerning EMC activities and related issues.

- Preparing reports, briefs, white papers, and working papers for the Committee to submit to other departments.

- Setting and implementing information service plans based on the Committee's needs.

- Preparing the final NEC Report.

- Collecting data and statistics.

- Providing the Committee with the studies and research required.

- Reporting candidates' expenditures in election campaigns.

- Managing legal affairs and auditing procedures.

2.4 Support services

- Setting and implementing policies, strategies, standards, and procedures for providing highly effective human resource services.

- Planning human resource requirements and preparing necessary budgets.

- Setting job plans and descriptions for human resources required for elections.

- Attracting and selecting competent staff (permanent/temporary/reserve).

- Registering staff and issuing ID cards.

- Training (staff in elections/ contact centres).

- Distributing staff and managing their affairs.

- Personnel affairs management during elections.

- Developing an appropriate method for calculating compensation of staff participating in administering elections.

2.4.1 Financial and administrative services

- Setting and implementing policies, strategies, standards, and procedures for providing highly effective administrative and financial services.

- Preparing budgets for the Election Management Committee, and monitoring budget execution.

- Performing purchase procedures, settling due amounts, and issuing checks.

- Making accounting entries in different original journals.

- Preparing the budget allocated for the committee, and for media campaigns, technological needs, office needs, logistic requirements, polling stations, hospitality, reception, transportation, accommodation and so on.

- Establishing compensation system for committee members/staff and organising elections/subcommittees.

- Setting and managing petty loans for officials at polling stations.

- Collecting fees (for registration/appeals).

- Conducting EMC-related public relations (PR) activities.

- Identifying the type of forms, applications, and reports specifically required and used in elections.

- Supplying polling stations with all required equipment.

- Providing accommodation and transportation for invited media representatives.

- Providing Meeting logistical needs and services for polling stations and media centres.

- Arranging official communications and correspondence with other regulatory bodies and units based on applicable powers.

- Managing archiving processes.

2.4.2 Human resources

- Setting and implementing policies, strategies, standards, and procedures for providing highly effective human resource services.

- Planning human resource requirements and preparing necessary budgets.

- Setting job plans and descriptions for human resources required for elections.

- Attracting and selecting competent staff (permanent/temporary/reserve).

- Registering staff and issuing ID cards.

- Training (staff in elections/contact centres).

- Distributing staff and managing their affairs.

- Personnel affairs management during elections.
- Developing an appropriate method for calculating compensation of staff participating in administering elections.

2.5 Media awareness

- Implementing the communication strategy as per the 'Guide for Media Handling of the Elections'.
- Implementing the draft plan prepared for the media and advertising campaigns.
- Carrying out different media-related activities and developing effective communication tools with relevant parties.
- Holding regular meetings with editors and journalists covering the elections to keep them updated.
- Supervising media centres, coordinating with mass media, and providing coverage materials, and bilingual media coordinators (fluent in Arabic and English).
- Training official spokesmen on how to deal with the media and relevant messages.
- Enacting regulations for communicating with media officials for election officers to abide by.
- Monitoring media coverage of the NEC, FNC Elections and political participation in the UAE, whether in local or global mass media.
- Distributing official election results through media outlets.
- Allocating special areas and seats for media representatives at every polling station.
- Informing Panel members of their electoral responsibilities, in coordination with the Electoral Processes Unit.
- Holding informative sessions about political participation, along with discussion panels and groups that cover the elections.

2.5.1 Contact centre

- Implementing the communication strategy as per the 'Guide for Media Handling of the Elections'.
- Implementing the draft plan prepared for the media and advertising campaigns.

- Carrying out different media-related activities and developing effective communication tools with relevant parties.

- Holding regular meetings with editors and journalists covering the elections to keep them updated.

- Supervising media centres, coordinating with mass media, and providing coverage materials, and bilingual media coordinators (fluent in Arabic and English).

- Training official spokesmen on how to deal with the media and relevant messages.

- Enacting regulations for communicating with media officials for election officers to abide by.

- Monitoring media coverage of the NEC, FNC Elections and political participation in the UAE, whether in local or global mass media.

- Distributing official election results through media outlets.

- Allocating special areas and seats for media representatives at every polling station.

- Informing Panel members of their electoral responsibilities, in coordination with the Electoral Processes Unit.

- Holding informative sessions about political participation, along with discussion panels and groups that cover the elections.

2.5.2 News

- Implementing the communication strategy as per the 'Guide for Media Handling of the Elections'.

- Implementing the draft plan prepared for the media and advertising campaigns.

- Carrying out different media-related activities and developing effective communication tools with relevant parties.

- Holding regular meetings with editors and journalists covering the elections to keep them updated.

- Supervising media centres, coordinating with mass media, and providing coverage materials, and bilingual media coordinators (fluent in Arabic and English).

- Training official spokesmen on how to deal with the media and relevant messages.

- Enacting regulations for communicating with media officials for election officers to abide by.

- Monitoring media coverage of the NEC, FNC Elections and political participation in the UAE, whether in local or global mass media.
- Distributing official election results through media outlets.
- Allocating special areas and seats for media representatives at every polling station.
- Informing Panel members of their electoral responsibilities, in coordination with the Electoral Processes Unit.
- Holding informative sessions about political participation, along with discussion panels and groups that cover the elections.

2.5.3 Media campaigns

- Implementing the communication strategy as per the 'Guide for Media Handling of the Elections'.
- Implementing the draft plan prepared for the media and advertising campaigns.
- Carrying out different media-related activities and developing effective communication tools with relevant parties.
- Holding regular meetings with editors and journalists covering the elections to keep them updated.
- Supervising media centres, coordinating with mass media, and providing coverage materials, and bilingual media coordinators (fluent in Arabic and English).
- Training official spokesmen on how to deal with the media and relevant messages.
- Enacting regulations for communicating with media officials for election officers to abide by.
- Monitoring media coverage of the NEC, FNC Elections and political participation in the UAE, whether in local or global mass media.
- Distributing official election results through media outlets.
- Allocating special areas and seats for media representatives at every polling station.
- Informing Panel members of their electoral responsibilities, in coordination with the Electoral Processes Unit.
- Holding informative sessions about political participation, along with discussion panels and groups that cover the elections.

3. The Emirates Committee

- Coordinating with the EMC about technical and administrative electoral issues in the Emirate, and more specifically:
- Locating the Committee office whilst maintaining communications with the EMC.
- Receiving the final list of the electoral panel and sending it to Committee members.
- Receiving electoral forms from the EMC and providing them in the Committee office.
- Coordinating with the Emirate's police to provide a sufficient number of police officers on election day, as per the election Security Committee guidelines.
- Collaborating with the local Municipality, jointly with the EMC.
- Recommending polling station location(s) in coordination with the EMC.
- Advising candidates of places for holding election rallies.
- Assessing the Emirate nominations against preset requirements before submission to the EMC.
- Assessing appeals for providing all information required before submission to the EMC.
- Monitoring compliance with election controls and measures and reporting any violations to the EMC.
- Performing any other functions within the authority of the Committee.

4. The Appeals Committee

- Drafting and implementing the appeals administration process.
- Reviewing and handling appeals concerning voter registration and the electoral roll.
- Reviewing and handling appeals concerning lists of candidates and their registration.
- Reviewing and handling appeals against voting results.

- Reviewing and handling appeals of any administrative violations during elections.
- Any other function within the authority of the Committee.

5. The Media Committee

- Raising public awareness of elections and encouraging their participation.
- Coordinating with different media channels to raise public awareness of the Federal National Council elections.
- Setting and implementing consolidated media policies, standards, strategies, and procedures (before, during, and after elections).
- Developing and implementing media programs and plans required for the elections.
- Developing a guide on how the media (whether visual, broadcast, or print media) covers and handles the elections.
- Coordinating the use of the official mass media in presenting candidate programs to ensure equal-opportunity exposure.
- Organising press conferences and maintaining media centres before, during, and after elections.
- Collaborating with professional media firms.
- Performing any other functions within the authority of the Committee.

6. The Security Committee

- Drafting and implementing security plan for elections.
- Developing security rules and criteria for polling stations to follow.
- Enacting consistent security measures for implementation at these polling stations.
- Identifying methods of dealing with election day issues and problems.
- Coordinating the provision and implementation of safety and security procedures and standards with every possible means.
- Any other function within the authority of the Committee.

Appendix 2:

A2.1. Requirements definition workshop

The business process requirements workshops spanned over four sessions and covered the following:

Session 1	
Title:	**Business Processes**
Objectives:	Agree on business processes. Demonstrate process mapping to existing e-Voting functionality, in order to identify gaps
Attendees:	All key business representatives; NEC's Legal advisor on Electoral Matters
Topics:	1. Introduction - Objective of Workshop and next steps 2. Processes • Before Election Day • During Election Day • After Election Day 3. As-is solution demo mapping to these processes 4. Regulatory and legal implications

Session 2	
Title:	**Solution Functionality and User Interfaces**
Objectives:	Review and agree on proposed functionality for key solution components and agree on content/functionality of user interfaces
Attendees:	Business Owners and End user representatives
Topics:	1. Details of functionality of each application 2. Basic wireframes to capture requirements related to screen content • Pollbook • Dashboard • Web Reporting • Voting Screens and Back office

Session 3	
Title:	Location, Roles and Responsibilities
Objectives:	Agree on high level requirements for different types of locations, understanding the constraints and roles required at each location. Identify exception handling and legal implications.
Attendees:	EIDA/NEC Technical Team; NEC Procurement; Election Organisers; NEC Legal Advisor on Electoral Matters

Topics:	Details of requirements and staff roles at each of the following locations
	    Data Centre
	    Election Operations Centre
	    Warehouse/testing site
	    Voting centers
	Non-functional requirements
	Exceptions handling processes
	Regulatory and legal implications

Session 4	
Title:	Information Exchange
Objectives:	Discuss and agree on information exchange requirements
Attendees:	EIDA/NEC Technical Team; EIDA/NEC staff who know about Voter and Candidate information and its format
Topics:	1. Voter validation/authentication • Integration with UAE ID Card 2. Information Exchange • Import Information: Voter list, Candidate names and pictures • Export Information: what information is to be exported from the system after election 3. Review BOM with details of equipment

These e-voting processes across the three stages are described here in terms of:

1. Business process diagrams which illustrate tasks, both manual and automated, and the roles involved in the execution of end-to-end electronic voting process, using the voting software.

2. Task description table that describes the various steps in the automated process.

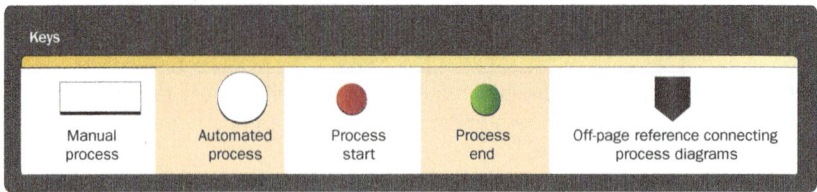

Figure A2.1 Key to symbols used in business process illustrations

Each section includes a business process diagram supported by a table listing process steps. Input, output, pre-requisites and steps for each key process are also described.

A2.2. Pre-election processes

After the system preparation and sealing steps have been completed, the election should be configured following the Election Configuration process.

Next we describe the steps required to accomplish this task.

A2.3. The election configuration process

Once the system is set up, the election configuration process is executed for each Emirate, in the presence of the Electoral and Admin board of that Emirate.

Input: for each Emirate: voters list, candidates list, voting rules, election timings.

Output: election configured and published.

Pre-requisites:

1. All information required to configure the election is provided by the National Election Committee in the desired format.

2. Electoral and admin boards for each Emirate are formed.

3. Staff for required roles are assigned and available.

A2.4. Election day process

Once the polling station is open, the voting process on election day proceeds in three stages.

A detailed process diagram for each stage:

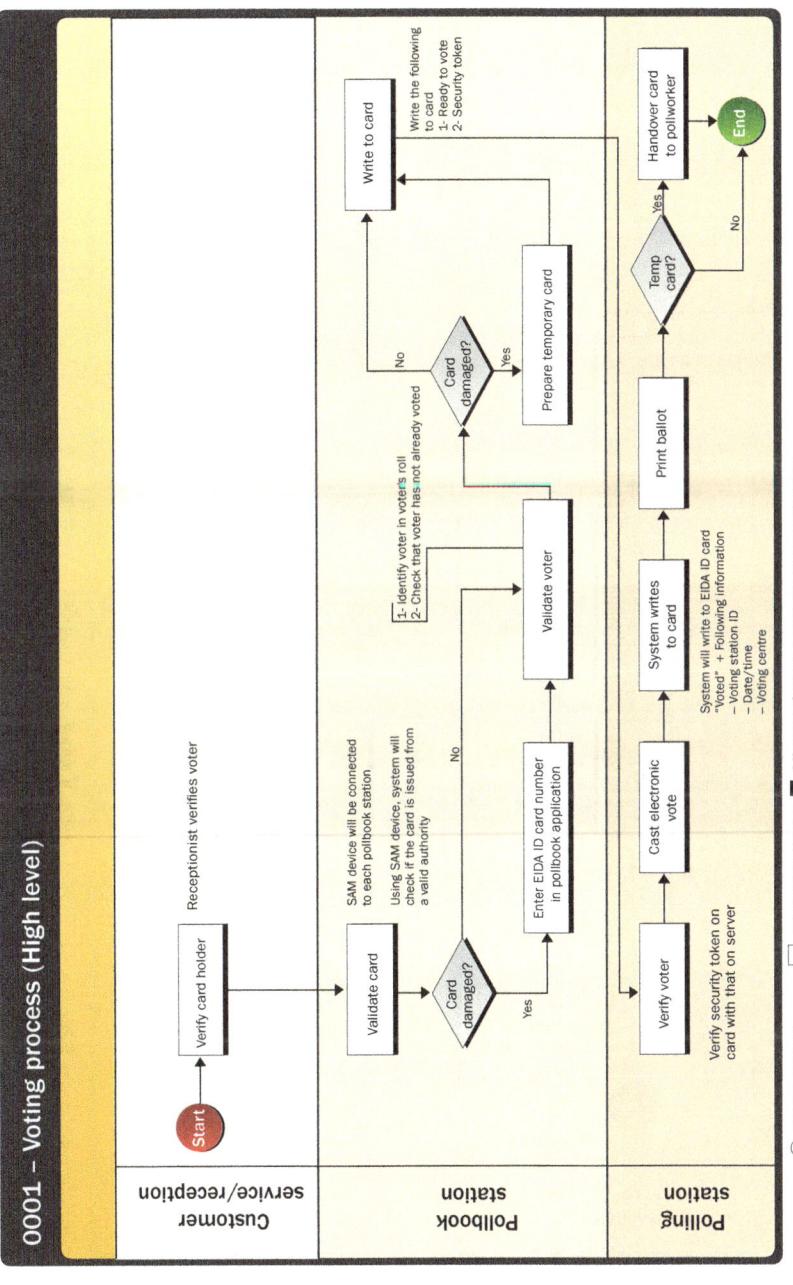

Figure A2.2 High level voting process

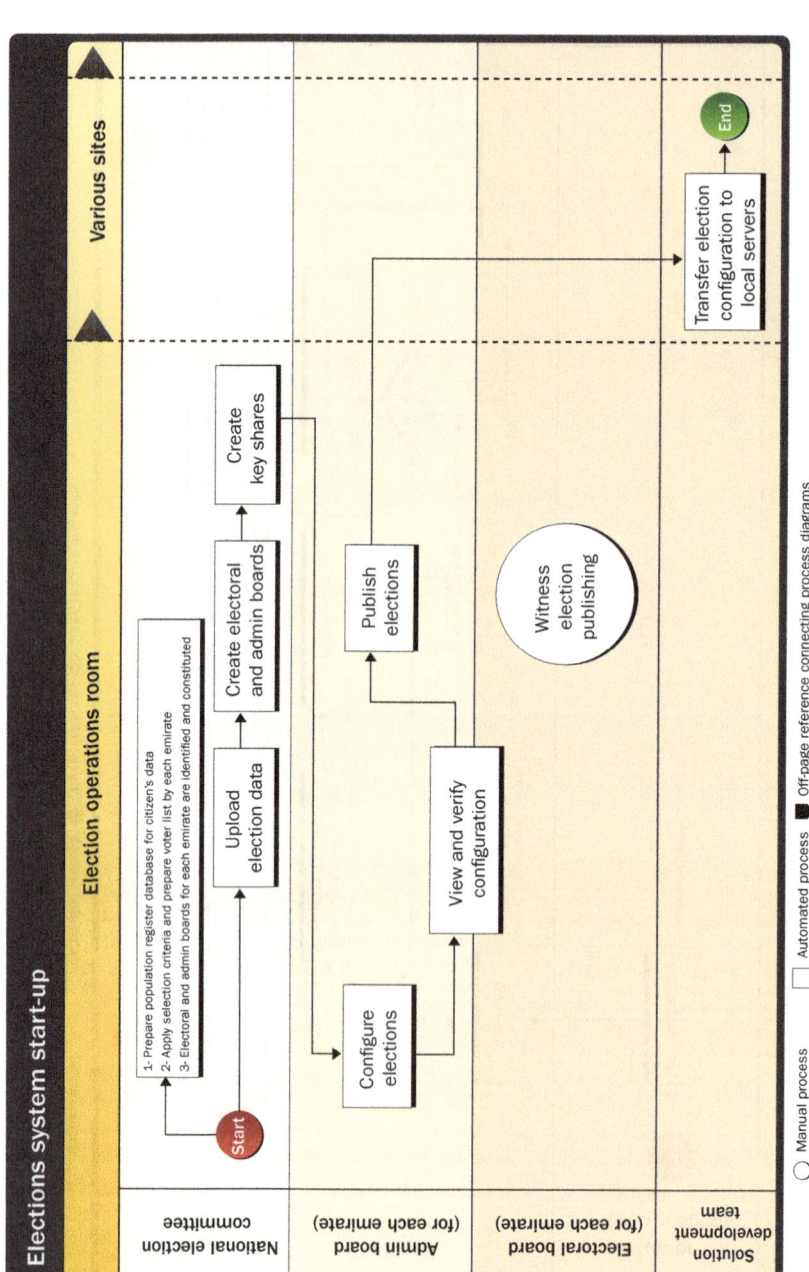

Figure A2.3 Election system start-up procedures

The diagram shows a process flow titled "Elections system start-up" with the following swimlanes:

Various sites / **Election operations room**

Roles (rows):
- National election committee
- Admin board (for each emirate)
- Electoral board (for each emirate)
- Solution development team

Process steps:
- Start
- Notes:
  1- Prepare population register database for citizen's data
  2- Apply selection criteria and prepare voter list by each emirate
  3- Electoral and admin boards for each emirate are identified and constituted
- Upload election data
- Create electoral and admin boards
- Create key shares
- Configure elections
- View and verify configuration
- Publish elections
- Witness election publishing
- Transfer election configuration to local servers
- End

Legend:
○ Manual process   □ Automated process   ▶ Off-page reference connecting process diagrams

Table A2.1   Pre-election technical process steps

Process Steps			
Steps	**Where**	**Who**	**Comments**
Software preparation			
1. Emirates ID provides population register database to each Emirate for preparing the electoral roll and finalising the list of eligible voters.	All seven emirates	Local authorities in each Emirate and the ruler's courts	
2. The required certificate authority (CA) for the election is created, and any related keys are also created.	EIDA PKI system	EIDA and solution provider	
3. The voting software is complied and built in a trusted environment in a suitable place.	Test site	Solution provider	
4. The built software is deployed: • In the data centre • In all the voting servers located in the test site • Voting terminals and ID verification units.	Several locations	Solution provider and election authority	
System sealing			
5. The servers in the data centre are logically sealed. • Database is physically and logically secured.	Several locations	EIDA	

A2.4.1. Election day procedure

The procedure for election day has been separated into two identified areas: polling stations and operations centres.

A2.4.1.1. The pre-opening process

Input: poll worker user credentials.

Output: voting terminals and servers are started and connected to central server.

Table A2.2   The election configuration process

Process steps			
Steps	**Where**	**Who**	**Comments**
Election configuration			
The election is configured • Election general data, according to reference manual. • The file(s) with candidates and pictures (associated to each Emirate) is uploaded in the voting system. • The file with the electoral roll (list of voters associated to each Emirate).	Operations centre	Business owner Auditor	Dependency: NEC to provide required files and information to configure the election in the agreed formats and times.
Each Electoral and Administration Boards are created in the correspondent operational centre server following a secret sharing scheme process: • The administration board is required to publish the election and to do some administration operations. • The electoral board holds the key required to decrypt the cast ballots and to sign election results.	Operations centre	Business owner Electoral board admin board	Dependency: Business owner to determine size and threshold for both boards. Both boards can be merged in one, but we strongly suggest to keep them separated to reduce the number of operations to be done by the potentially high profile persons that will compose the electoral board. The admin board is usually composed of more technical persons than the electoral board trust to delegate such tasks.

Table A2.2   The election configuration process

Process steps			
Steps	Where	Who	Comments
The election is published, i.e. its configuration is generated and exported so it can be deployed in the other voting servers of each Emirate: • Any stakeholder can review the configured data (candidates, dates, and so on). • The admin board signs the exported data in the publishing process.	Operations centre	Business owner Auditor Admin board	• Deployment on each voting centre server by each Emirate.
Each electoral and administration boards are created in the correspondent operational center server following a secret sharing scheme process: • The administration board is required to publish the election and to do some administration operations. • The electoral board holds the key required to decrypt the cast ballots and to sign election results.	Operations centre	Business owner Electoral board admin board	Dependency: Business owner to determine size and threshold for both boards. • Both boards can be merged in one, but we strongly suggest to keep them separated to reduce the number of operations to be done by the potentially high profile persons that will compose the electoral board. • The admin board is usually composed of more technical persons than the electoral board trust to delegate such tasks.

| Table A2.2 | The election configuration process |

Process Steps			
Steps	Where	Who	Comments
Election configuration			
The election is published, i.e. its configuration is generated and exported so it can be deployed in the other voting servers of each Emirate: • Any stakeholder can review the configured data (candidates, dates, and so on). • The admin board signs the exported data in the publishing process.	Operations centre	Business owner Auditor Admin board	Deployment on each polling station server by each Emirate.
The configuration files generated on each operational centre server are transferred to the data centre servers: • They are uploaded and deployed on each polling station servers of the corresponding Emirate. • They are uploaded and deployed on each instance of the solution of the corresponding Emirate.	Test site	Business owner Auditor Admin board	**Exceptions:** **Issue:** No connectivity to data centre. **Solution:** take the DVD manually to data centre and testing site.
Final deployment testing			
Deployment is validated and each decentralised voting server is sent to its destination polling station.	Several sites	System integrator	Additional logical sealing can be done on the directories containing election configuration files.
Final Deployment Testing is performed: • Connectivity from polling stations' voting terminals and pollbooks. • Dashboard.	Each polling station Operations centre	System integrator	

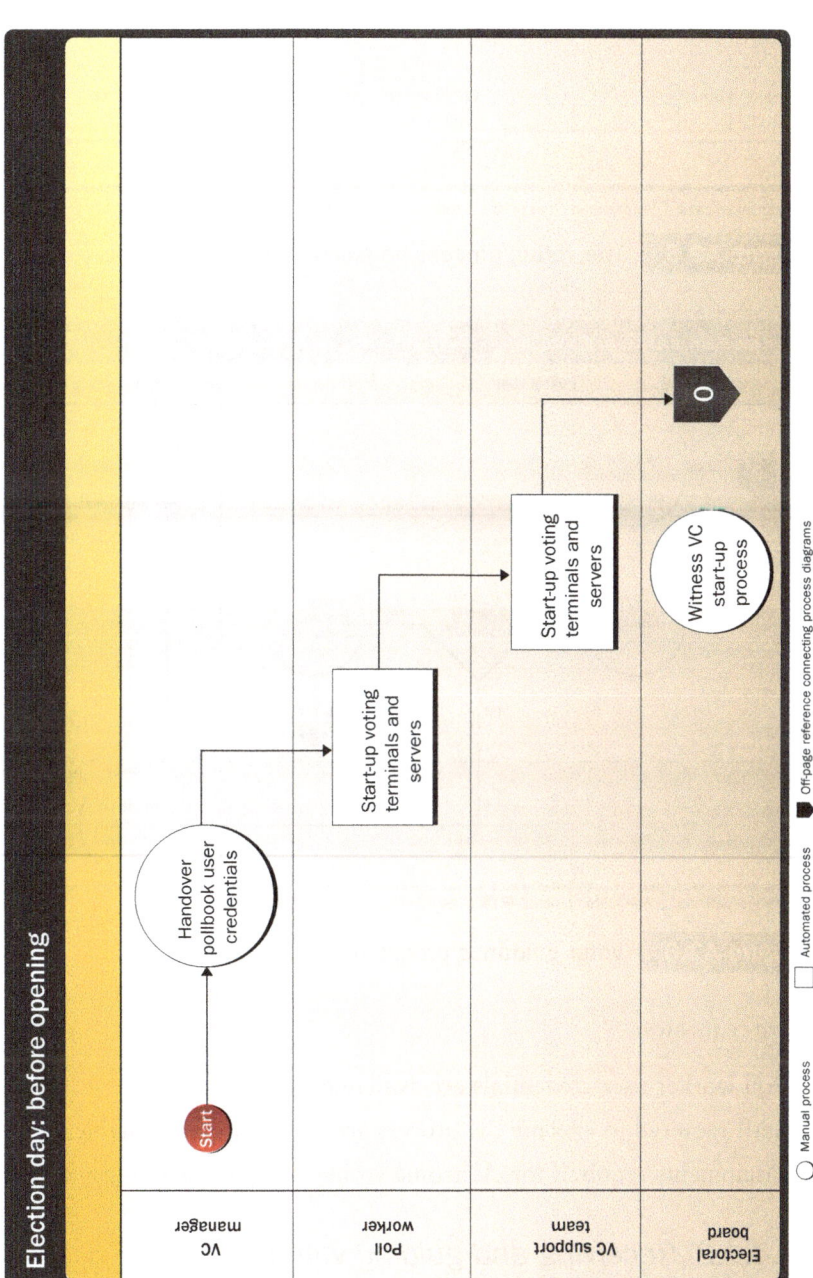

Election day: before opening

VC manager	Poll worker	VC support team	Electoral board
Start → Handover pollbook user credentials	Start-up voting terminals and servers	Start-up voting terminals and servers	Witness VC start-up process

○ Manual process   ☐ Automated process   ◤ Off-page reference connecting process diagrams

Figure A2.4   Business process diagram (election date)

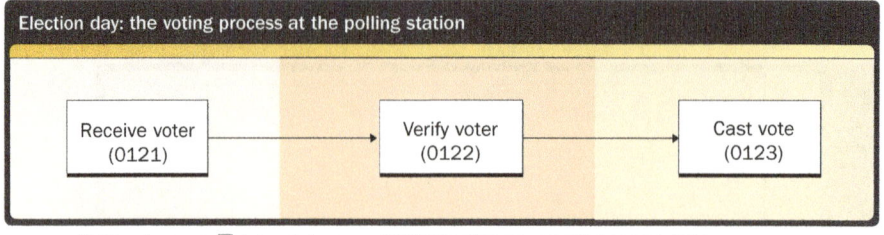

Figure A2.5 The voting process on election day

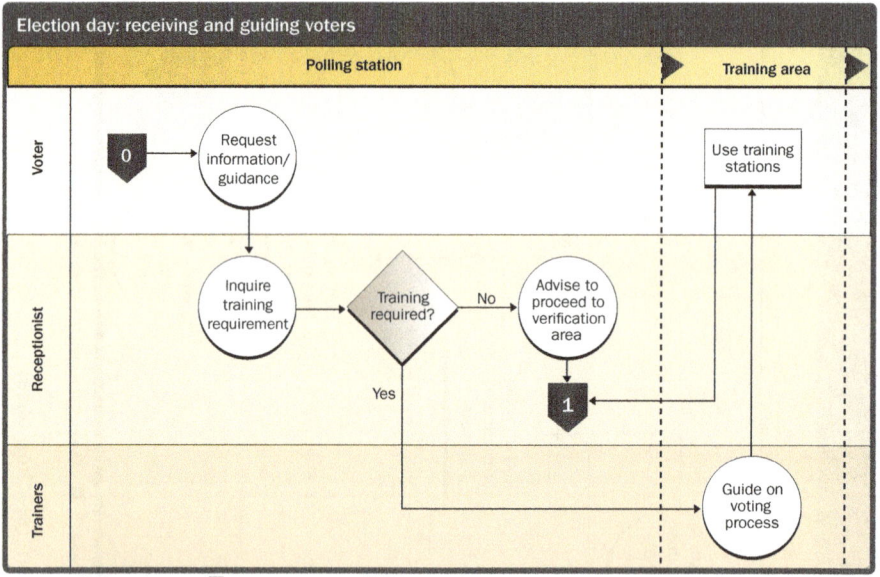

Figure A2.6 Voter guidance procedures

Pre-requisites:

1. Poll worker user credentials are available.

2. Staff required to execute the process are assigned and available.

3. Equipments required for electronic voting is deployed and tested.

A2.4.1.2. Receiving and guiding voters

This process is carried out by polling station receptionists and does not involve the use of any automated system.

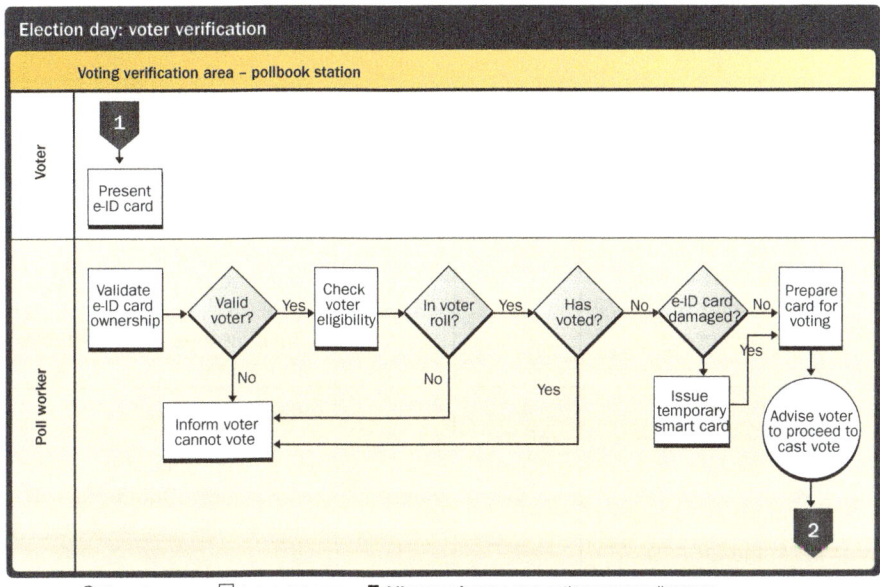

O Manual process   ☐ Automated process   ♥ Off-page reference connecting process diagrams

Figure A2.7   Voter verification process

The receptionists are the first point of contact for the voter and do the following:

1. Check if the voter has the Emirates ID required for voting.

2. Guide the voter to the training area if requested to do so.

3. Guide the voter to the verification area.

4. Provide any information on the voting procedure if asked.

Table A2.3   Centre pre opening check process

Process steps			
Steps	**Where**	**Who**	**Comments**
Pre-opening check			
Before the polling station is opened to voters, poll workers start up all the computers, etc. and check again the connectivity.	Polling stations	Poll workers Polling station support team	▪ Mock voting and/or zero vote verification could be done. ▪ Electoral board need to be present at the moment of initiating the process. ▪ Candidates or representative could be present as the electoral rule allow it.

Table A2.4 The voter authentication process

Process steps			
Steps	**Where**	**Who**	**Comments**
Voter authentication			
A voter gets into the polling station and is addressed to one poll worker managing one pollbook.	Polling stations	Poll workers Voters	• Reception area provide customer service if required as first point of contact. • Training area also available if required.
The poll worker validates the identity and eligibility of the voter: • Checks the picture in the EIDA ID. • In some cases, uses biometric authentication. • Inserts the EIDA ID in the pollbook reader. • The pollbook automatically checks the eligibility of the voter (he is in the electoral roll of this Emirate and has not voted before).	Polling stations	Poll workers Voters	**Exceptions** **Issue:** What to do if the voter has not EIDA ID (but other ID document like a passport)? **Solution:** No vote allowed. **Issue:** What to do if no connectivity to data centre? **Solution:** Vote will proceed. Vote to be stored locally until connectivity resumes. Then, votes stores locally will be synchronised with DC DB. Note: A special page showing connectivity logs/ graphics, coupled with some dynamic green/red light showing on the header of the pollbook application will be available.

Table A2.4 The voter authentication process (*Cont'd*)

Process steps			
Steps	Where	Who	Comments
Issue temporary card			
If the voter has EIDA ID which is not readable (e.g. broken card), a blank card will be generated with a valid credential to vote: • The poll worker searches for the voter in the pollbook by typing his name and last name. • The pollbook will tell whether the voter is able to vote or not. • If the voter is eligible, the poll worker will issue a blank card to allow the voter to vote.	Polling stations	Poll workers Voters	• Pollbook functioning as credential issuing point.

A2.4.1.3. The voter verification process

Input: Emirates ID card

Output:

1. The voter is verified.

2. The Emirates ID card is verified and prepared for voting.

Pre-requisites:

1. The voter must have an Emirates ID card.

2. The staff is required to carry out the procedure is available.

Table A2.5 The e-ballot vote casting procedure

Process Steps			
Steps	Where	Who	Comments
Voter cast e-ballot			
6. The voter accesses a voting terminal and logs in: • Inserts his EIDA ID in the terminal reader. • Touches the screen to activate the process.	Polling stations	Voters	
7. The voter makes his selections if the authentication is correct. • The terminal displays the candidates related to the voter's Emirate. • The voter makes his selections. • The voter clearly sees which candidate are the selected options. • The voter presses the 'continue' button.	Polling stations	Voters	• Selection from 0 to the max number of candidates by Emirate. • Blank voting is allowed. • Under voting is allowed. • Over-vote is not allowed.
8. The terminal displays the voter a confirmation screen with the selected candidates: • The voter can go back to the previous screen and change his selection (previous step). • The voter confirms the selections by pressing the button 'cast ballot'. • The ballot is encrypted in the voting terminal and sent to the voting servers.	Polling stations	Voters	• Only selected candidates (with name and photo) will be displayed.

Table A2.5   The e-ballot vote casting procedure (*Cont'd*)

Process Steps			
Steps	**Where**	**Who**	**Comments**
Printing of paper copy			
9. The terminal shows confirmation of the correct storage of the ballot and instructions to the voter about the next steps: ▪ Wait until the digital ballot copy with selection is printed. ▪ Take the ballot copy. ▪ Voter confirms a valid ballot printing. ▪ Remove the e-ID card.	Polling stations	Voters	**Exceptions:** The printer does not print the piece of paper. **Solution:** the voter requests the help of a poll worker, who can access a special page and request a new printing by typing a special password.
10. The voting terminal returns to the initial stage waiting for a new voter. ▪ The voter is directed by poll workers to deposit his ballot copy in the ballot box. ▪ This ballot box must be different to the one used to store any potential paper ballots cast. ▪ The voter return the smart card in case it was issued by the poll worker.	Polling stations	Poll workers Voters	Special paper must be used for printing ballots which would allow to detect fake ones.
Voter finishes			
11. The voter leaves the polling station.	Polling stations	Voters	Business owners to consider surveying voters about their experience.

Table A2.6 The monitoring dashboard

Process steps			
Steps	Where	Who	Comments
Monitoring			
12. Check the participation data in the dashboard.	Operations centre	Business owner System integrator Auditor	▪ Information displayed by Emirate.
13. Check the status of the polling stations (whether they are connected or not).	Operations centre	System integrator Auditor Business owner	▪ Each VC will periodically perform a service check against the data centre. ▪ Dashboard will have an historic and graphics of each VC connectivity.
14. Check the availability/ performance of the data centre equipment.		Data centre technicians	▪ This is responsibility of the provider of the data centre. ▪ Probably it could be resolved by reports stating every two hours the status of the servers and the bandwidth.

A2.4.1.4. The vote casting procedure

Input: smart card with voter's credentials.

Output: printed copy of ballot to be placed in ballot box.

Pre-requisites:

1. The voter is eligible to vote.
2. The e-voting system is available.

A2.4.1.5. In the operations centre

This is an outline of actions that should be carried out on election day in the operations centres.

| Table A2.7 | The procedure for closing elections and generating the results |

Process steps			
Steps	**Where**	**Who**	**Comments**
End of voting period			
15. At the configured time (date/time), the voting system will stop accepting ballots. Also, poll workers will not allow to access any voter to the polling station.	Polling stations	Poll workers	**Exceptions** **Issue:** voters still in the polling stations waiting to vote when closing time happens. **Solution:** (1) If the Emirate has more than one VC, then configure the voting system to automatically stop a few hours after the designated election closing time (e.g. 1–2 hours), and close the process manually when the polling stations report no more voters are present. (2) For those Emirates with only one polling stations in the Emirate then Election Admin Board can extend the 'Election closing time at their discretion'.
16. The poll workers will shut down all voting terminals and poll-books (log out and switch them off). ▪ Operations centre will be notified that the polling stations is 'closed' for voting.	Polling stations	Poll workers	

Table A2.7	The procedure for closing elections and generating the results (Cont'd)

Process steps			
Steps	Where	Who	Comments
Validate synchronisation			
17. The designated technician/poll worker will check in the local servers that all the ballots are synchronised with the central servers. • Notify operations centre of the checking result. • Follow agreed procedure if some ballots are pending.	Polling stations	System integrator Poll worker	**Exception** **Issue:** ballots pending synchronisation **Solution:** some options available: 1. Using a back up connection to finish synch. 2. Exporting the data to a DVD and submit it manually to the operations centre.
Paper ballot counting			
18. If paper ballot is allowed, poll workers will proceed to count paper ballots. • Break seal of ballot boxes. • Manually count the paper ballots. • Fill out report with results, usually signed by several poll workers. • Any other documentation and procedures associated with paper ballot handling will be followed.	Polling stations	Poll workers	Dependency: • NEC to define the procedure to manage/store paper ballots. • The form will be similar to the one used in polling station and will be blank to be filled by an electoral officer. • After that this will be put into the dashboard application.

Table A2.7	The procedure for closing elections and generating the results (*Cont'd*)

Process steps			
Steps	**Where**	**Who**	**Comments**
19. Local register of results to the Emirate. operations centre. ▪ A representative of the polling station will call the Emirate operations centre to report the results (in cases where the polling station does not include an operations centre. ▪ An operator in the operations centre will record the results into the system.	Polling stations Operations centre	Poll workers Operators	▪ It will be the dashboard where paper votes will be recorded and consolidated them with the results of the electronic votes.
Decryption and tabulation of e-votes			
20. When notified by all polling stations that they are ready (step above), and using a secure connection from the operations centre, the electoral board in presence of the candidates attending the event, will authorise and initiate the online mixing and tallying process.	Operations centre	System integrator Auditor Business Owner	**Exceptions** **Issue:** no connectivity with the data centre. **Solution:** Mixing and tallying process could be performed against local data base, since synchronisation process keep all votes locally for contingency.

Table A2.7 The procedure for closing elections and generating the results (*Cont'd*)

Process Steps			
Steps	**Where**	**Who**	**Comments**
21. The mixing/ tabulation process can start: • The electoral board members insert, one by one, their cards to reconstruct the decryption key. • The ballots are shuffled and randomly decrypted following a mixing cryptographic protocol. • The results tabulated are displayed in the mixing server (number of votes per candidate). • Also, files with the decrypted ballots are exported from the server to be displayed.	Operations centre	Business owner System integrator Auditor Electoral board	The results are also displayed in the operations room of each Emirate.
Results dissemination			
22. The results are uploaded in the WEB result application for its public dissemination. These results shall include data from paper counting and electronic counting.	Operations centre	Business owner System integrator Auditor	The final results could be made public after the election (i.e. accessible by all the citizens) We can import the elections and candidates into the dashboard using the CSV export file in Pnyx, so that we also have all the election configuration in the dashboard.

Table A2.7	The procedure for closing elections and generating the results (*Cont'd*)

Process steps			
Steps	**Where**	**Who**	**Comments**
Final auditing			
23. Audit servers • The logical sealing of the different servers can be validated to see that they remain as before sealing. • Other logs can be validated too.	Operations centre Data centre?	Business owner System integrator Auditor	If VPN is open after elections to the data centre, all validations can be done from the operations centre. This process can also be done before the mixing process, but it takes time and will delay the publishing of the results.

Note: A documented procedure will be executed at each polling station to close the election by not allowing any more votes to be cast including turning off the voting machines or stop the Pnyx server (side effect: synchronisation to the data centre could still not be finished).

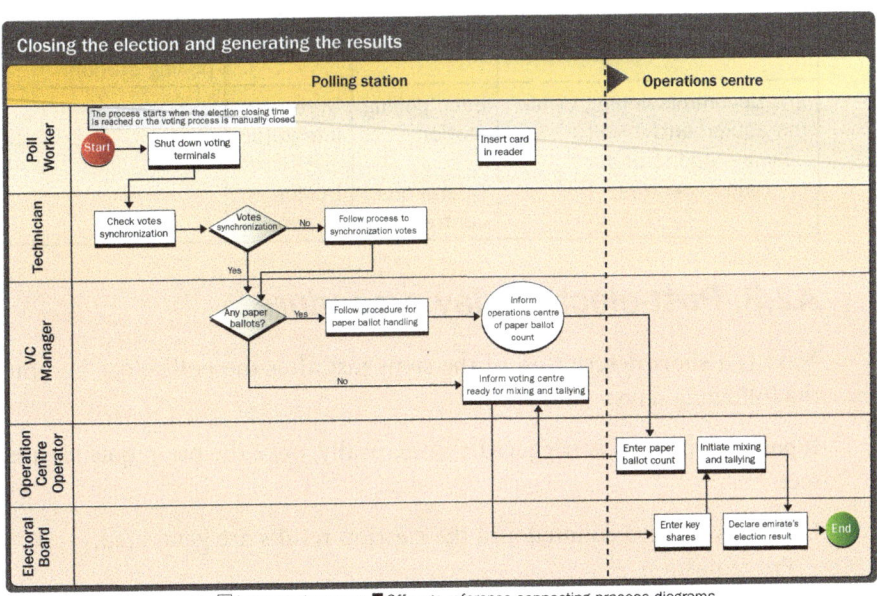

Figure A2.8	Closing the elections and generating the results

Table A2.8 Decommissioning of equipment

Process Steps			
Steps	Where	Who	Comments
24. Archiving the election data • The required data will be copied in WORM media (e.g. DVD) and delivered to NEC for its custody until the period open to claims is closed. This data should allow the repetition of a mixing/tabulation process if required. After this period expires, the data can be destroyed. • Data in the servers related to ballots and voters should be removed.	Several sites	Business owner System integrator Auditor	Pending to define which data is to be stored. Usually logs, ciphered ballots and mixing output, plus all configuration files and the electoral board and administration cards.
Archiving and decomissioning			
25. Audit paper copies of ballots, if required.	Polling stations	Poll workers Business owner	Manual process, and probably only done on a fraction of the ballots and/ or in a single polling station.
26. Decommissioning of all the equipment: polling stations and operations centres.	Each polling station Operations centre	System integrator Business owner	

A2.3. Post-election day procedure

Next is a short description of the steps just after the polls close and on the following days.

Input: The process is triggered automatically, once the pre-requisites are met.

Output: votes are counted and the election results are generated.
   Pre-requisites:

1. The election is closed manually by the polling station manager or the closing time is reached.

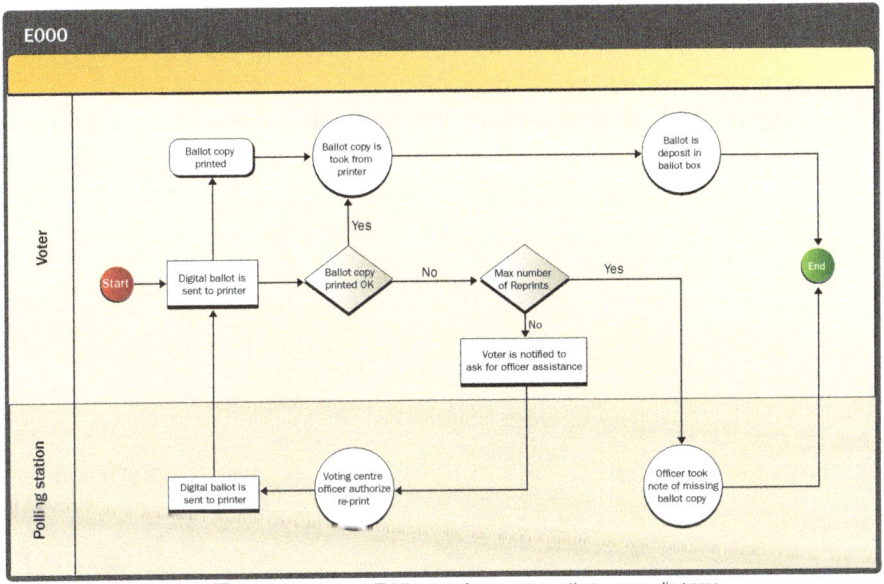

○ Manual process ☐ Automated process ◗ Off-page reference connecting process diagrams

Figure A2.9 | Exceptional ballot copy printing procedure

A2.3.2. Decommissioning equipment

Equipment at the polling station must be decommissioned and moved to designated sites, once the elections are declared complete and closed.

Input: instructions and authorisation from the NEC to start decommissioning.

Output: equipment decommissioned.

Pre-requisites:

1. All the information on the server is backed up and secured in the designated area.

A2.4. Contingencies for business procedure exceptions

A2.4.1. Ballot printing fails

Ballot copies are required to be printed and stored in the ballot box for auditing purposes.

If the printer fails, the voter should request assistance from the polling station officer to authorise the reprinting process.

Only three attempts will be authorised by the system.

Figure A2.9 is a procedure diagram representing this exceptional process.

Part 3
Identity management

Identity and mobility in a digital world

Abstract: Mobile identity management has attracted the attention of both the public and private sectors in the last few years. In the context of service delivery, modern mobile communication networks offer more convenient approaches to developing citizen-centric applications. However, taking into consideration the need for compelling user authentication and identification, secure communication in mobile environments remains a challenging matter. This article explores the potential role of government-issued smart identity cards in leveraging and enabling a more trusted mobile communication base. It delves into the identity management infrastructure program in the United Arab Emirates (UAE) and how the smart identity card and overall system architecture have been designed to enable trusted and secure transactions for both physical and virtual mobile communications.

Keywords: *Identity management; mobile identity; national authentication infrastructure; wireless PKI, identity provider; UAE.*

1. Introduction

Now more than ever, the world is transitioning to the digital sphere, in which the best possible use is made of digital technologies. Our transition to the digital world has been rapid and innovative, and it is now shifting us towards a more converged existence. It was not too long ago that discrete devices worked in isolation, but now they are ubiquitous and provide us with a seamless experience in service delivery that can be accessed virtually from anywhere and at anytime. For example, one could submit a service request on the Web, follow and track the delivery progress using a landline on IVR, call a contact centre and update the request, provide additional information at a kiosk, pay using a credit card from a mobile phone, and receive the physical goods at the designated location and a confirmation of delivery via SMS. This is the reality of the digital world.

In this digital domain, all facets of mobile identity management are gaining wider attention from both the public and private sectors. Their significance lies in their contribution towards providing trust for transactions in user-centric applications and their impact on the overall context of service delivery (1). However, this is not as tranquil as it may sound. The primary challenge is that mobile identity management systems need to be multi-laterally secure and allow appropriate user access to data, while enabling privacy and anonymity (2).

Referring to the specificity of the mobile services field in modern networks, Srirama et al. (2006) state that there are some 'characteristics unique to the mobile paradigm, the increased complexity of emerging handheld devices, the greater sensitivity to security and load related problems in wireless infrastructure and increased complexities of scale' (3). This implies that we first need to have unified identification criteria that allow us to identify mobile and virtual individuals. The role of an 'identity provider' is crucial to confirming the credibility of the parties participating in a service or transaction. It should provide sufficient credentials for service providers' to explicitly authenticate mobile users, thus enabling trust in the transactions being made securely.

In this article, we examine the role that a national identity infrastructure could play in the facilitation of mobile environments, with specific reference to the United Arab Emirates (UAE). In light of the current shortcomings in the existing literature about government practices in the field, we attempt to create government-published content to support the current body of knowledge that could support the development of both the practical and research fields.

This article is structured as follows: Section 2 provides a short overview of the role of identity, in light of converging digital technologies; Section 3 introduces the UAE's digital identity management infrastructure; Section 4 briefly describes the UAE's identity system and its service eco-system; Section 5 provides an overview of the digital ID profile and how it helps in establishing trust in e-transactions; Section 6 presents how the UAE identity management system supports mobility, both from individual and enterprise perspectives; Section 7 looks at some government plans that attempt to integrate mobile phones with its national identity management infrastructure to improve the accessibility of its services; and finally, Section 8 concludes with a discussion.

2. The convergence of digital technology

Digital convergence is an evolving reality. A key point is that this convergence is not just limited to technology, but also reaches out to user experiences. User interfaces, information, and services are all converging into computer-mediated systems that are independent of the communication channel, device or tool (see Figure 6.1). The user is not tied down geographically and may interact, communicate, collaborate, and share information in many new and different ways.

Unquestionably, the body (entity) of central focus in digital environments is the person seeking the service, or in other words, the transaction initiator. Different stakeholders in this transaction lend their support, and the service provider delivers the requested service and ensures service fulfillment. A more revolutionary form of business operations may appear when an irrefutable identity is made available for the service seeker, who is actually anonymous in the digital world.

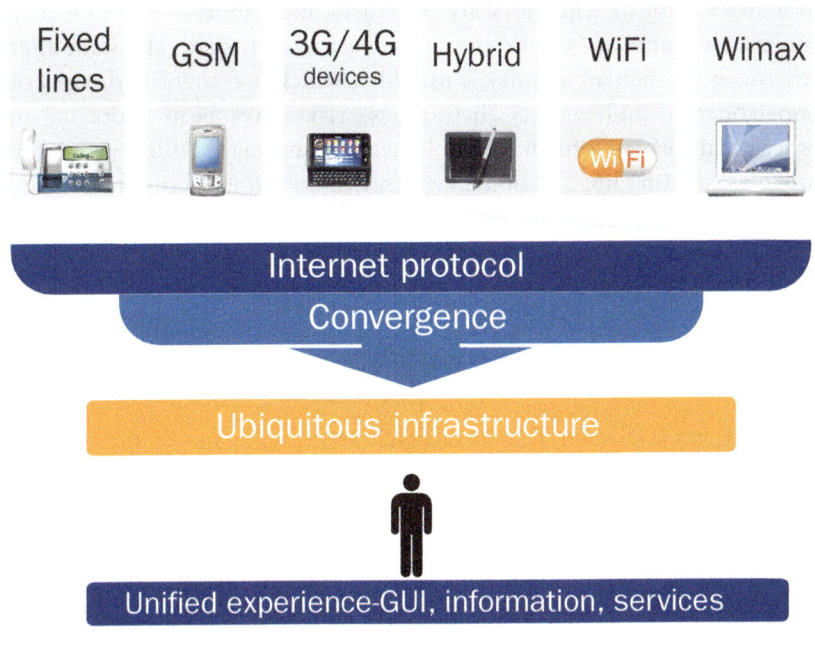

Figure 6.1 Convergence of digital technologies and citizen interfaces

Our digital world spreads across several areas, but they all meet in their needs for identification. In other words, the underlying enabler to our digital world is identity, which is:

- Uniform across multiple channels of communication.

- Standardised and usable in all contexts of identity verification.

- Issued by a trusted identity provider that enables authentication on demand across various communication channels.

In reality, there are multiple providers of identity for any given entity, all jostling for space, which enable cross verification and authentication over diverse services. Thus, as long as the services are virtual, inter-dependent identity verification can suffice. Yet, with real services like physical goods, secure communication enablement, and financial transactions, higher trust requirements are needed for the completion of the transactions.

With such needs in mind and in an attempt to build the infrastructure for digital economies, many governments worldwide have awoken to the need to provide trusted and ubiquitous identities to their citizens (4, 5, 6). In a space with multiple identity providers, government-issued identity credentials stand to become the most trusted construct. The issue that remains with such an identity is usability, which is fraught with risks of impersonation and identity theft. These risks have been addressed in other countries, as some have implemented, and many others are in the process of setting up, a national authentication infrastructure to provide seamless identification that bridges multiple service channels to enhance users' experiences and enable secure digital transactions (7).

3. The UAE digital identity system

The UAE has set a clear vision for digital identity issuance in the country. Figure 6.2 depicts the smart identity card issued by the UAE to all of its citizens and residents. This comes as part of its national identity management program (also referred to as national identity management infrastructure) that was launched in mid-2005. In the seven short years since its launch, UAE has been a leading country in the Middle East and Africa in issuing more than 8.5 million digital certificates to its population. This represents 96 per cent of its total population; 99.9 per cent of the

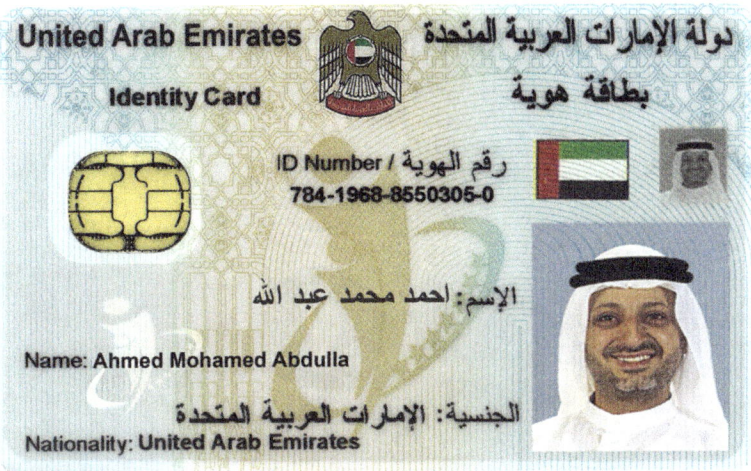

Figure 6.2 The UAE smart identity card

citizens and 95 per cent of the expatriate resident population. The digital identity provided by the UAE is composed of a set of credentials delivered in the form of a smart card, which includes a unique national identification number, biometric data (fingerprints), and a pair of PKI digital certificates; one for authentication and the other for signature (8).

4. The UAE ID and service eco-system

The secure credentials issued by the UAE national identity management infrastructure are designed to support their use by the citizens and residents in both physical and digital environments. Smart identity cards with digital credentials are provided to facilitate government and public sector service delivery transactions, from across manned counters to transactions online. This is supported by multi-factor authentication capabilities and advanced identity verification mechanisms. Figure 6.3 shows the context of the national identity card and the service eco-system in the UAE.

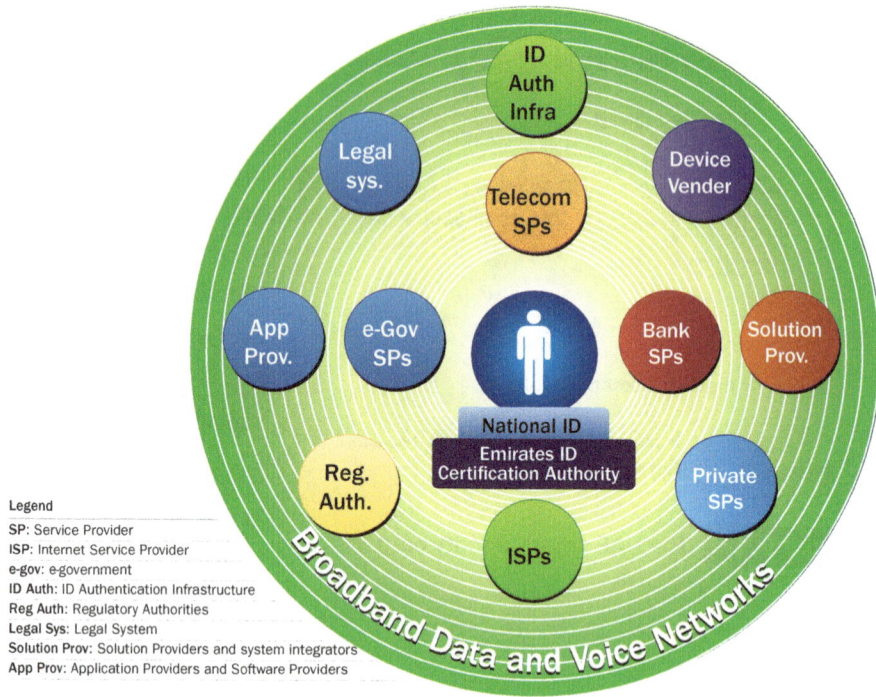

Legend
SP: Service Provider
ISP: Internet Service Provider
e-gov: e-government
ID Auth: ID Authentication Infrastructure
Reg Auth: Regulatory Authorities
Legal Sys: Legal System
Solution Prov: Solution Providers and system integrators
App Prov: Application Providers and Software Providers

Figure 6.3   Identity and service eco-system

5. The UAE national ID card and the establishment of trust

The UAE national identity card is a smart combi-card with both contact and contactless communication capabilities. The card is an instrument that is packaged to carry the physical identity details of the card holder, along with the digital ID profile, in the smart chip. The digital ID profile consists of:

1. A unique national identity number (IDN).

2. Biometrics (fingerprints).

3. A pair of digital certificates issued from the population certification authority (CA) of the public key infrastructure set up for this purpose.

Whenever a card is presented in any electronic (remote) transaction, the online validation centre (i.e., national authentication centre) validates the card by verifying its authenticity and the expiry date of the card. Whenever personal data is read from the card, the data is presented with the Emirates Identity Authority's digital signature confirming that the data on the card has not been tampered with. That the card holder is who he/she claims to be is established using multiple factors of authentication, including a PIN verification and biometric verification.

A software development kit (SDK) is available on the government's Internet portal to promote usage and integration of the identity card with service providers' systems in both the government and public sectors. An applet is also available online for card holders to download onto their personal computers to register themselves and use certain e-services that require strong authentication. These are but a few examples of the identity card's uses for which the SDK and the card applet have supported the development of auto-ID and mobile applications. Thus, the UAE national ID card establishes trust in the virtual, digital world and enables online transactions.

6. Enabling mobile transactions

It is important to note here that the term mobility transcends individuals, enterprises, and the government. Each entity attempts to reach out to its respective stakeholders and be 'available' at all times and over different communications media. Let us look at mobility from both individual and enterprise perspectives.

6.1. Individual mobility

As mentioned earlier, individuals in the UAE are provided with smart cards that carry trusted identity credentials issued by the government. The digital identity associated with verifiable credentials is now available in the card holder's wallet. This enables the individuals to present their IDs and credentials on demand to gain access to government or private sector services. Thus, the ID card provides verifiable credentials for a person wherever he/she goes.

6.2. *Enterprise mobility*

Since verifiable credentials are available on demand, enterprises can now deliver services at the citizens' convenience without the citizens having to physically visit the service providers' premises. For instance, service delivery agents carrying mobile PDAs could verify the service recipient's identity by reading the identity card data stored in the chip. (see Figure 6.4).

Further authentication requirements could be met using PIN verification and/or biometric verification; both in contact as well as contactless mode. The current handheld terminals have add-on options for smart card readers that enable card reading through contact capabilities for example. The contactless capabilities of the identity card on the other hand enable the NFC mode to communicate with the HHTs/PDAs.

Cryptographic enhancements for biometric authentication on the HHTs/PDAs are possible by means of security access modules (SAMs), in the form of SIMs. Alternatively, the authentication infrastructure provided allows the HHTs to communicate securely on 3G/4G networks and to establish the necessary crypto-environment for biometric authentication.

The match-on-card feature for biometric authentication is used in both contact and contactless modes. This feature opens numerous possibilities for enabling e-commerce, in which identity is presented in a virtual world and goods are delivered to real entities that are identified

Figure 6.4   Mobile card reader terminal

authentically. A leading bank in the UAE is currently in the process of setting up a pilot system with the national identity cards for biometric authentication that uses handheld terminals to deliver bank debit/credit cards to their rightful owners.

Another initiative being developed is a pilot for opening customer accounts using customer services officers in banks to provide customer data by reading identity cards and signing to open the account by using the digital certificates in the identity card. This will follow the use of the identity card to conduct secure online financial transactions, which will use digital signatures and timestamp capabilities. This, too, takes the banking business to the living rooms of their customers. The UAE's telecom service providers are also coordinating to use the national identity cards with NFC smart phones and enable mobile contactless transactions.

7. The UAE national ID card and the future of mobility

The UAE identity management infrastructure has laid a strong foundation and framework for identity verification and identity authentication. It has contributed to the development of an enhanced technological environment that facilitates secure transactions with true mobility. Developments in telecommunication networks and smart phones are revolutionising people's lives. Figure 6.5 provides an overview of mobile subscribers worldwide. The UAE has nearly 150 per cent mobile penetration, and smart phones are at nearly 100 per cent utilisation. The government is planning to integrate mobiles phones with its national identity management infrastructure, in order to improve the accessibility of its services.

Public Key Infrastructure is an integral part of the UAE identity management system, with its population certification authority (CA) providing digital certificates for identity verification and digital signing. This capability can be extended to provide derived credentials for mobile users. This is a very interesting capability, since this infrastructure can be used to provide a wireless PKI. In other words, the UAE government is working to issue certificates, in conjunction with the national identity, to mobile subscribers. The telecom networks would facilitate the transport and installation of the certificates onto their subscribers' phones. The private keys would be secured in an encrypted location on the SIM card

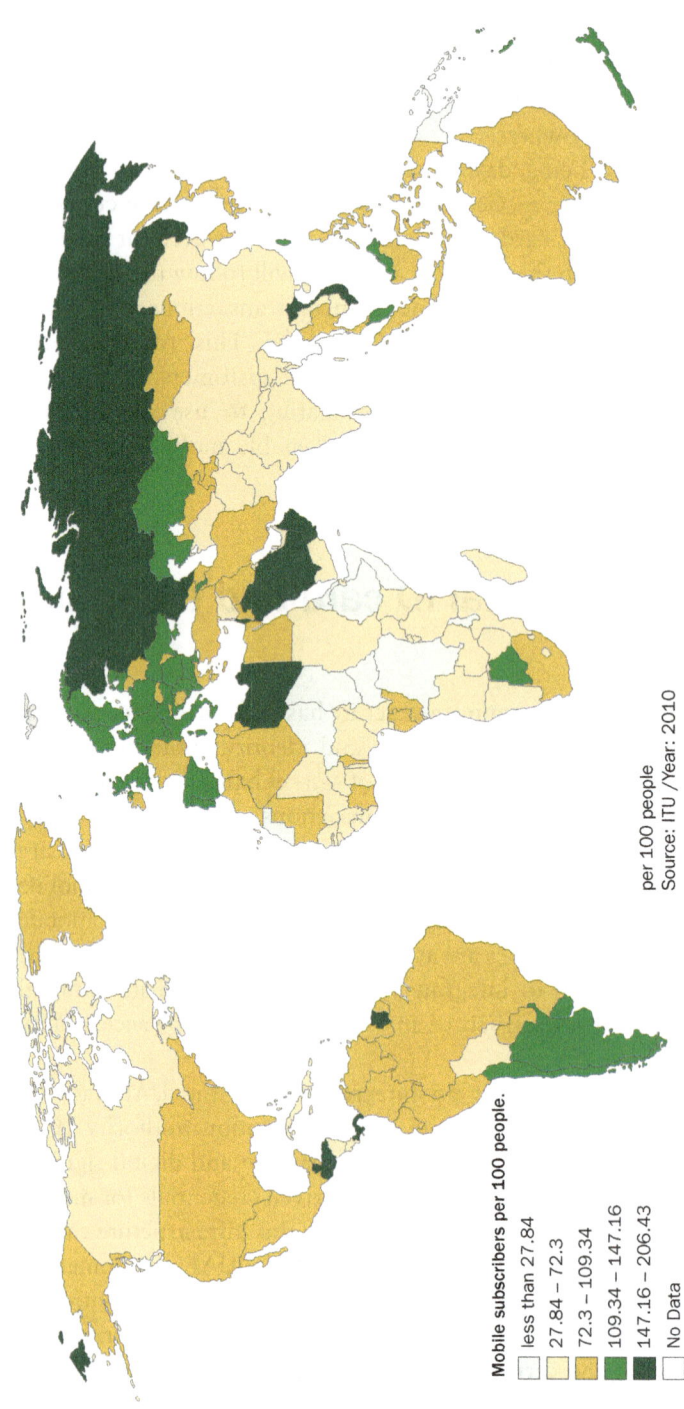

Mobile subscribers per 100 people.

- less than 27.84
- 27.84 – 72.3
- 72.3 – 109.34
- 109.34 – 147.16
- 147.16 – 206.43
- No Data

per 100 people
Source: ITU /Year: 2010

Figure 6.5 Number of mobile subscribers by country, per 100 people

itself, and not in the operating system of the mobile phones. This would tie in the mobile ID credentials to authentic national ID card holders and valid phone subscribers.

This will serve two major objectives for the UAE. First, it would provide the ability for all national identity card holders to conduct secure transactions with different service providers using their mobile phones in a manner similar to the presence of the identity card. Digital signatures with time stamping would greatly increase productivity and e-government transactions, and it would enable secure government communications.

Second, it would bring all non-identity card holders under the umbrella of the national identity program. Thus, all visitors subscribing to mobile services would have digital certificates issued by the UAE national identity management infrastructure, which would enable them to conduct secure mobile transactions. This would also contribute to national security.

A typical example of using the mobile ID credentials derived from the UAE national ID card is depicted below in Figure 6.6.

The UAE has other exciting developments in the making that will use the post-issuance scenarios of the identity card to provide additional mobile services. One of the key initiatives in this direction is the ability of the national identity infrastructure to provide OTP (one-time-password)

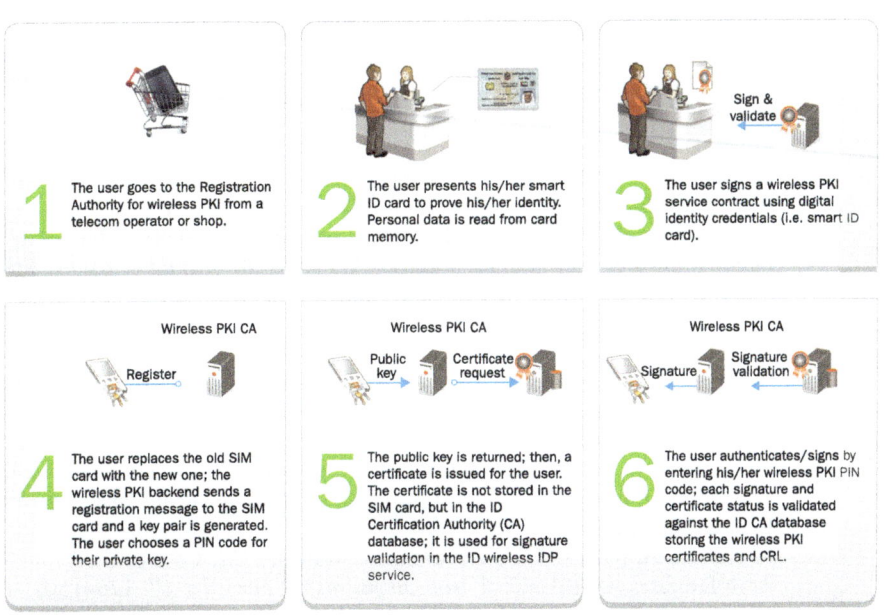

1 The user goes to the Registration Authority for wireless PKI from a telecom operator or shop.

2 The user presents his/her smart ID card to prove his/her identity. Personal data is read from card memory.

3 The user signs a wireless PKI service contract using digital identity credentials (i.e. smart ID card).

4 The user replaces the old SIM card with the new one; the wireless PKI backend sends a registration message to the SIM card and a key pair is generated. The user chooses a PIN code for their private key.

5 The public key is returned; then, a certificate is issued for the user. The certificate is not stored in the SIM card, but in the ID Certification Authority (CA) database; it is used for signature validation in the ID wireless IDP service.

6 The user authenticates/signs by entering his/her wireless PKI PIN code; each signature and certificate status is validated against the ID CA database storing the wireless PKI certificates and CRL.

Figure 6.6 Wireless PKI registration and use

functions using Open AuTHentication (OATH) standards. As a post-issuance service, national identity cards could be loaded with an OTP applet that could then work in conjunction with NFC-enabled phones for an OTP for secure banking and financial transactions.

8. Conclusion

Most of the initiatives outlined here are likely to be instigated by other governments elsewhere in the world. Although some countries have initiated similar pilot projects, these are mainly being driven by the private sector, with little involvement from government bodies. The development of mobile networks and communication technologies will push governments to rethink their 'role' and definition of 'identity' in the digital world.

Governments are facing increasing pressure to improve the quality of life and address the changing and ever-complex needs of their populations. There will be no choice but to accept the concept of mobile identification. The future of 'mobile identification' in the information era will determine the competiveness of countries and their readiness to survive the challenges of tomorrow. Let us wait and see.

References

1. Royer, D. (2012) 'Economic Aspects of Mobility and Identity'. *Future of Identity in the Information Society.* Available at: *http://www.fidis.net/fileadmin/fidis/deliverables/fidis-wp11-del11.3.economic_aspects.pdf*
2. Deuker, A. and Royer, D. (2009) 'Next Generation Networks'. *Future of Identity in the Information Society.* Available at: *http://www.fidis.net/fileadmin/fidis/deliverables/new_deliverables2/fidis-wp11-del11_11_Next_Generation_Networks.final.pdf*
3. Srirama, S., Jarke, M. and Prinz, W. (2006) 'Mobile Host: A Feasibility Analysis of Mobile Web Service Provisioning'. *Proceedings of 4th International Workshop (CAiSE'06) on Ubiquitous Mobile Information and Collaboration Systems*: 942–953. Luxembourg, June 5–6.
4. Al-Khouri, A. M. (2012) 'An Innovative Approach for e-Government Transformation'. *International Journal of Managing Value and Supply Chains* 2 (1): 22–43.
5. Al-Khouri, A. M. (2012) 'e-Government Strategies The Case of the United Arab Emirates (UAE)' (sic). *European Journal of ePractice* 17: 126–150.

6. Al-Khouri, A. M. (2012) 'Emerging Markets and Digital Economy: Building Trust in The Virtual World'. *International Journal of Innovation in the Digital Economy* 3 (2): 57–69.

7. Al-Khouri, A. M. (2007) 'Electronic Government in the GCC Countries'. *International Journal of Social Sciences* 1 (2): 83–98.

8. Al-Khouri, A. M. (2012) 'PKI in Government Digital Identity Management Systems'. *European Journal of ePractice* 14: 4–21.

9. Haddon, L. (2003) 'Domestication and Mobile Telephony'. In Katz, J. E. (ed), *Machines that Become Us: The Social Context of Personal Communication Technology*: 43–55. New Brunswick, NJ: Transaction Publishers.

10. Keeney, J., Lewis, D., O'Sullivan, D., Roelens, A., Wade, V., Boran, A. and Richardson, R. (2006) 'Runtime Semantic Interoperability for Gathering Ontology-Based Network Context'. In Network Operations and Management Symposium, 2006. NOMS 2006. 10th IEEE/IFIP: 56–65.

11. Kim, W. (2005) 'On Digital Convergence and Challenges'. *Journal of Object Technology* 4 (4): 67–71. Available at: *http://www.jot.fm/issues/issue_2005_05/column5.pdf*

12. Mantena, R. and Sundararajan, A. (2004) 'Competing in Markets with Digital Convergence'. Available at: *www.web-docs.stern.nyu.edu/old_web/emplibrary/04-12Sundararajan.pdf*

13. Mueller, M. (1999) 'Digital Convergence and its Consequences'. *The Public* 6: 11–28. Available at: *http://javnost-thepublic.org/article/pdf/1999/3/2/*

14. Narendra, S. (2012) 'Connecting Identity and Mobility: A Secure, Scalable and Sustainable Mobile Wallet Approach'. *IQT Quarterly* 4 (1): 18–21. Available at: *http://tyfone.com/IQT_Quarterly_Summer2012_Tyfone_article.pdf*

15. Strassner, J., Foghlu, M. O., Donnelly, W. and Agoulmine, N. (2007) 'Beyond the Knowledge Plane: An Inference Plane to Support the Next Generation Internet'. In: Global Information Infrastructure Symposium, 2007, First International GIIS: 112–119.

Data ownership:
who owns 'my data'?

Abstract: The amount of data in our world today is substantial. Many of the personal and non-personal aspects of our day-to-day activities are aggregated and stored as data by both businesses and governments. The increasing data captured through multimedia, social media, and the Internet are a phenomenon that needs to be properly examined. In this article, we explore this topic and analyse the term data ownership. We aim to raise awareness and trigger a debate for policy-makers with regard to data ownership and the need to improve existing data protection, privacy laws, and legislation at both national and international levels.

Keywords: *Data ownership; big data; data protection, privacy*

1. Introduction

Organisations today have more data than ever. Advancements in technology play a critical role in generating large volumes of data. According to a study published by Information Week, the average company's data volumes nearly double every 12 to 18 months (Babcock, 2006). Databases are not only getting bigger, but they also are becoming real time (Anderson, 2011; Sing et al., 2010).

Evolving integration technologies and processing power have provided organisations with the ability to create more sophisticated and in-depth individual profiles based on one's online and offline behaviours. The data generated from such systems are increasingly monitored, recorded, and stored in various forms in the name of enabling a more seamless customer experience (Banerjee et al., 2011; Halevi and Moed, 2012; Rajagopal, 2011).

The subject of who actually 'owns' the data or, in other words, the term 'data ownership' has attracted the attention of researchers in the past few years. Data transmitted or generated through digital communication channels becomes a potential for surveillance. Data ownership issues are thus likely to proliferate. For instance, Facebook's famous announcement that users cannot delete their data from Facebook caused a furore, and Mark Zuckerberg (one of five co-founders of Facebook) was equally famous in his response: '...it's complicated'.

Indeed it is! In today's interconnected world driven by the Internet, powered by gigabyte network operators, we leave a significant and by no means subtle data trail. The often-asked question, and the issue of discussion today is: Who owns this data? In order to answer this question, it is important for us to step back to examine the very nature of what we call 'data'.

2. Data: a matter of interpretation

There much confusion about what 'data' really is in today's world. The truth is that data is no more than a set of characters, which, unless seen in the context of usage, has no meaning (Wigan, 1992). Data is what one uses to provide some information. The context and the usage provide a meaning to the data that constitutes information. Thus, data in stand-alone mode has no relevance and therefore no value. When there is no value in data, then one would surmise that ownership is not an issue. That is the paradox of data ownership. Figure 7.1 illustrates a data value pyramid developed by Accenture. The pyramid has three levels, starting from raw data, up to the insights and then the transaction levels. The base of the pyramid features raw, less differentiated, and thus less valuable data. Moving up the pyramid creates larger value and revenue opportunities.

As such, governments and public sector institutions consider data a public utility (WEF, 2011). They tend to label our personal data as 'corporate data' and argue that without it, they cannot function (Holloway, 1988). It is no wonder that the volume of stored data in today's organisations has increased exponentially.

It is in this context that the ownership of data needs to be considered. As data is generated, it is stored. When we speak about data ownership, we refer to the storage process. If so, then the ownership of data storage resides with the owner of the storage. Thus, we as individuals, the

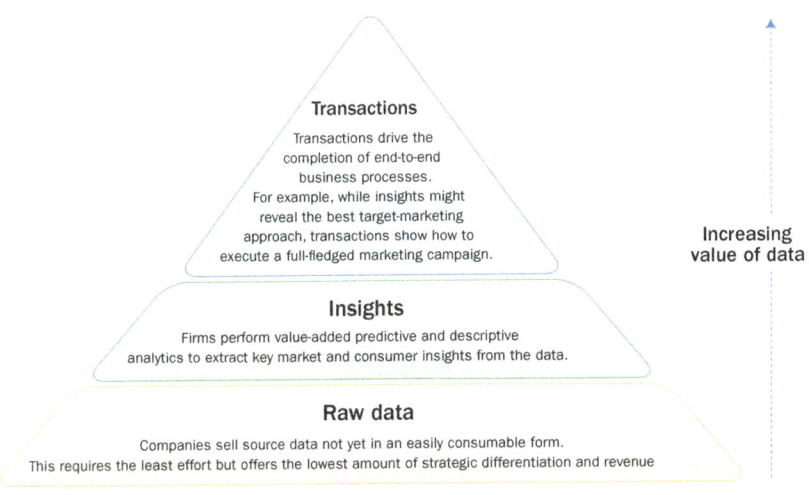

Figure 7.1 Data value pyramid

government as our governing agent, law enforcement agencies and the courts, security agencies, our service providers, and our network operators who enable us to move our data are all our data owners. They own the storage systems and thereby the data held within such systems. In addition, the emergence of customer data integration (CDI) and of master data management (MDM) technologies has enabled the integration of disparate data from across multiple silos into commonly defined, reconciled information accessible by a range of systems and business users (Dyché, 2007).

We would like to pause here and examine the concept of 'my data'. Just what is 'my data'? Do we consider information from friends and family that we hold to be 'my data'? Do bank statements and credit statements sent by banks qualify as 'my data'? Would the financial statement sent by a company to me as a shareholder qualify as 'my data'?

We would argue no. 'My data' in its strict sense, comprises just our personal attributes. This is the data that I own. I use 'my data' as information to identify myself for my personal gain, whether physical, logical, or emotional. 'My data' is thus in the open and either implicitly or explicitly shared. When I share the data, I delegate the ownership. Thus, 'my data' has multiple owners, and the number of owners increases with each share. Figure 7.2 provides an illustration of this viewpoint.

As the number of transactions increases with 'my attributed', 'my data' grows and, in turn, increases the number of data owners. What

Figure 7.2   Potential owners of 'my data'

essentially is happening is that with every transaction, information is shared as data. Each time information is shared, new data is generated. As new data is generated, new ownership is created. Figure 7.2 illustrates just a tip of the proverbial iceberg of data generation.

3. The personal data ecosystem and ownership

Typically, organisations can capture different personal data in a variety of ways (Marc et al., 2010):

- Data can be 'volunteered' by individuals when they explicitly share information about themselves through electronic media, for example, when someone creates a social network profile or enters credit card information for online purchases.

- 'Observed' data is captured by recording users' activities (in contrast to data they volunteer). Examples include Internet-browsing preferences, location data when using cell phones, or telephone usage behaviour.

- Organisations can also discern 'inferred' data from individuals, based on the analysis of personal data. For instance, credit scores can be calculated based on a number of factors relevant to an individual's financial history.

Each type of personal data (see Figure 7.3), volunteered, observed, or inferred, can be created by multiple sources (devices, software applications), stored, and aggregated by various providers (online retailers, Internet search engines, or utility companies), and then analysed for a variety of purposes for many different users (end users, businesses, public organisations).

So, in all this chaos of data generation and delegated ownership, where does true ownership lie? The answer to this question is found in truth, veracity, and therefore verifiability. Each time information is generated, a set of data related to this information is created. This data, when relayed further, should stand up to scrutiny and verification. The contention is that the source that can verify this data and confirm the veracity of the information is the 'true owner' of the data. Figure 7.4 presents a conceptual model to illustrate the source of truth and data ownership.

Even in the context of complex online transactions, this statement would remain valid. Intimately linked to the growth of 'big data' are such technological trends as the growth of mobile technology and wireless devices, the emergence of self-service channels, the broad adoption of cloud-based services, and the expansion of social networking and remote collaboration (SAP, 2012). For instance, Google is synonymous with its search engine and provides a host of services that require a user to log in. When a logged-in user searches the Web, data is generated related to the user's search patterns. While the search information itself does not belong to Google, the data collected on the search patterns do. Any analysis based on these search patterns can be traced back to Google's search data.

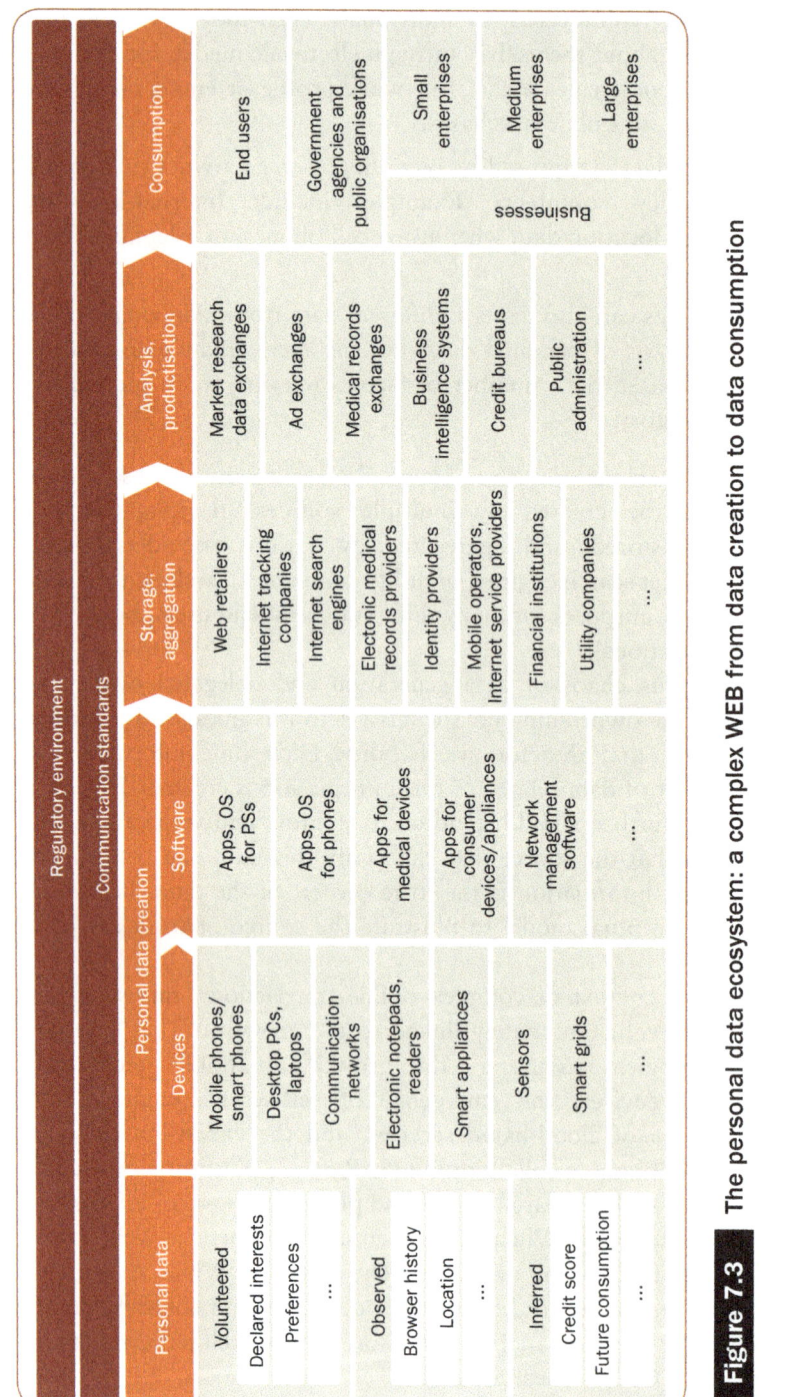

Figure 7.3 The personal data ecosystem: a complex WEB from data creation to data consumption

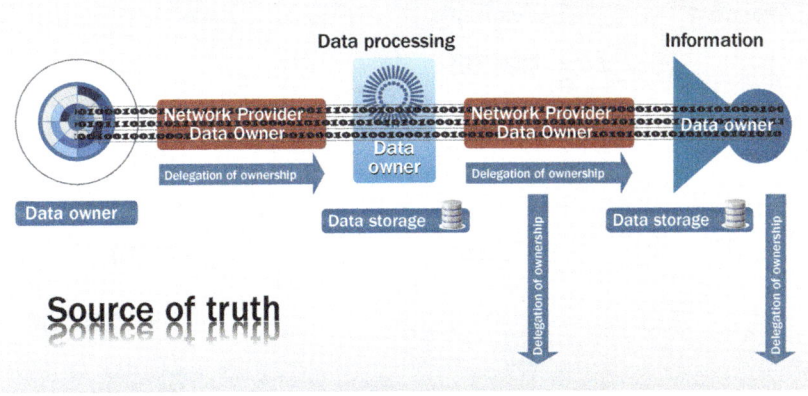

Source of truth

Figure 7.4 Source of the truth of data ownership

As depicted in Figure 7.5, 'big data' companies collect and analyse massive amounts of data under the argument that they can spot trends and offer users niche insights that help create value and innovation much more rapidly than conventional methods. This generates more data, the analysis of which is, more often than not, as useful as the original information. This analytical information now belongs to the person and/ or the organisation that performed the analysis. This brings up the critical issue of data usage and information usage.

For example, in a well-publicised incident that occurred in August 2006, America Online published a dataset of search results. This data was collected from the searches conducted by users and was intended to provide analytical material to researchers. The data published was anonymous, without any reference to the users who carried out the searches. The searchers' identities were distinguished as numbers. Five days later, however, The New York Times was able to locate one of those searchers by linking her search history to other public data, such as the phonebook (Barbaro and Zeller, 2006).

4. The need to redefine the ecosystem

So who possesses the right to use the information and data that truly does not belong to oneself? This is an issue that transcends borders of

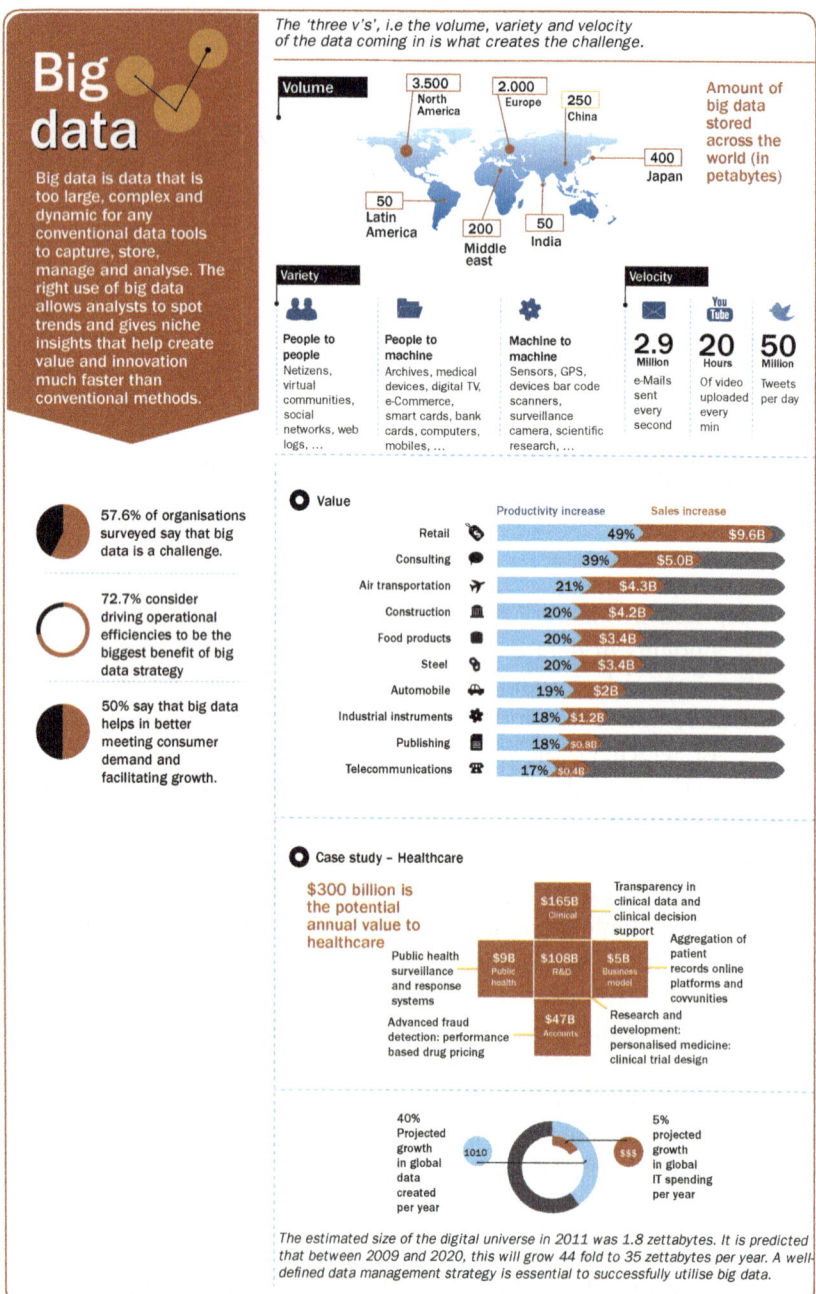

The 'three v's', i.e the volume, variety and velocity of the data coming in is what creates the challenge.

Big data

Big data is data that is too large, complex and dynamic for any conventional data tools to capture, store, manage and analyse. The right use of big data allows analysts to spot trends and gives niche insights that help create value and innovation much faster than conventional methods.

Volume

3.500 North America
2.000 Europe
250 China
400 Japan
50 Latin America
200 Middle east
50 India

Amount of big data stored across the world (in petabytes)

Variety

People to people
Netizens, virtual communities, social networks, web logs, ...

People to machine
Archives, medical devices, digital TV, e-Commerce, smart cards, bank cards, computers, mobiles, ...

Machine to machine
Sensors, GPS, devices bar code scanners, surveillance camera, scientific research, ...

Velocity

2.9 Million
e-Mails sent every second

20 Hours
Of video uploaded every min

50 Million
Tweets per day

57.6% of organisations surveyed say that big data is a challenge.

72.7% consider driving operational efficiencies to be the biggest benefit of big data strategy

50% say that big data helps in better meeting consumer demand and facilitating growth.

Value

	Productivity increase	Sales increase
Retail	49%	$9.6B
Consulting	39%	$5.0B
Air transportation	21%	$4.3B
Construction	20%	$4.2B
Food products	20%	$3.4B
Steel	20%	$3.4B
Automobile	19%	$2B
Industrial instruments	18%	$1.2B
Publishing	18%	$0.8B
Telecommunications	17%	$0.4B

Case study – Healthcare

$300 billion is the potential annual value to healthcare

$165B Clinical
$9B Public health
$108B R&D
$5B Business model
$47B Accounts

Public health surveillance and response systems

Advanced fraud detection: performance based drug pricing

Transparency in clinical data and clinical decision support

Aggregation of patient records online platforms and covvunities

Research and development: personalised medicine: clinical trial design

40% Projected growth in global data created per year
1010
5% projected growth in global IT spending per year
$$$

The estimated size of the digital universe in 2011 was 1.8 zettabytes. It is predicted that between 2009 and 2020, this will grow 44 fold to 35 zettabytes per year. A well-defined data management strategy is essential to successfully utilise big data.

Figure 7.5 Big data

commerce, ethics, and morals, leading to privacy issues and the protection of privacy. It is trivial that the current personal data ecosystem is fragmented and inefficient (WEF, 2011). For many participants, the risks and liabilities exceed the economic returns. On the other hand, personal privacy concerns are inadequately addressed, and current technologies and laws fall short of providing the legal and technical infrastructure needed to support a well-functioning digital infrastructure (ibid.). Instead, they represent a patchwork of solutions for collecting and using personal data in support of different institutional aims, and subject to different jurisdictional rules and regulatory contexts (e.g., personal data systems related to banking have different purposes and applicable laws than those developed for the telecom and healthcare sectors).

It is of importance that governments play a more active regulatory role in modernising their existing policy frameworks to protect personal data from the unlawful processing of any data (Robinson et al., 2009). The government should move away from a regulatory framework that measures the adequacy of data-processing by measuring compliance with certain formalities, and towards a framework that instead requires certain fundamental principles to be respected, and that has the ability, legal authority, and conviction to impose harsh sanctions when these principles are violated (ibid.).

A recent report published by the World Economic Forum recommended that the personal data ecosystem be debated and redefined (WEF, 2012). This prompts all stakeholders to come to a consensus on some key areas, including the security and protection of data, development of accountability, and agreements on principles or rules for the trusted and allowed flow of data in different contexts (see Table 7.1).

These principles should be global in scope, but also applicable across sectors and focused beyond merely minimising data collection, storage, and usage of data to protect privacy. The principles need to be built on the understanding that to create value, data must move, and moving data requires the trust of all stakeholders. Organisations will need to develop and implement a comprehensive data governance program that should be based on these guiding principles. This should help organisations to design and implement more comprehensive structures and to put in place solid accountability that altogether establishes a coordinated response to key issues of trust, transparency, control, and value.

| Table 7.1 | Key principles to guide the development of the personal data ecosystem |

Guiding Principle	Description
Accountability	Organisations need to be held accountable for appropriate security mechanisms designed to prevent theft and unauthorised access of personal data, as well as for using data in a way that is consistent with agreed upon rules and permissions. They need to have the benefit of 'safe harbour' treatment and insulation from open-ended liability, when they can demonstrate compliance with objectively testable rules that hold them to account.
Enforcement	Mechanisms need to be established to ensure organisations are held accountable for these obligations through a combination of incentives, and where appropriate, financial and other penalties, in addition to legislative, regulatory, judicial, or other enforcement mechanisms.
Data permission	Permission for usage needs to be flexible and dynamic to reflect the necessary context and to enable value-creating uses, while weeding out harmful uses. Permission also needs to reflect that many stakeholders–including but not limited to individuals–have certain rights to use data.
Balanced stakeholder roles	Principles need to reflect the importance of rights and responsibilities for the usage of personal data and strike a balance between the different stakeholders–the individual, the organisation, and society. They also need to reflect the changing role of the individual from a passive data subject to an active stakeholder and creator of data. One perspective that is gathering momentum, though it is far from being universally accepted, is that a new balance needs to be struck that features the individual at the centre of the flow of personal data, with other stakeholders adapting to positions of interacting with people in a much more consensual, fulfilling manner.
Anonymity and identity	The principles need to reflect the importance of individuals being able to engage in activities online anonymously, while at the same time establishing mechanisms for individuals to effectively authenticate their identity in different contexts, so as to facilitate trust and commerce online.
Shared data commons	The principles should reflect and preserve the value to society from the sharing and analysis of anonymised datasets as a collective resource.

5. Conclusion

The term 'data ownership' is likely to attract more attention from both practice and research fields. The private sector will continue to use data as a source of competition and growth. Advocators will always justify their practices that this contributes to productivity, innovation, and competitiveness of entire sectors and economies (Manyika et al., 2011). Governments will need to play a more active role to protect citizens' privacy rights, in the light of the evolving world we live in today.

Governments will inevitably need to redesign and enforce data protection privacy laws and legislation. This will require establishing policies at both national and international levels. As such, governments will need to open up dialogue to establish comprehensive data protection and privacy laws that could be implemented globally. This should be followed by a clearly articulated set of standards, policies, procedures, and responsibilities regarding data ownership and data-related activities that may minimise any detrimental outcomes in the event of a data breach (PTACT, 2010). Governments should also focus on enforcing transparency. Public education programs might be a good initiative to support understanding of how individuals can protect their personal data, and how such data are being stored and used (Manyika et al., 2011).

As time passes, we are likely to see increasing public concern about privacy and trust in today's interconnected online environments. Governments will need to help the public to understand where they should position themselves within this spectrum. There will be challenging times for governments to keep up with the pace of technology development, and those lagging behind will have a hard time indeed.

References

1. Alstyne, M. V., Brynjolfsson, E. and Madnick, S. E. (1994) 'Why not One Big Database? Principles for Data Ownership'. Available at: *http://dspace. mit.edu/bitstream/handle/1721.1/2516/SWP-3695-31204002-CISL-94-03. pdf?sequence=1*

2. Anderson, J. H. (2011) 'Real Time Systems Resource Management'. *Real Time Systems* 47 (5): 387–388.

3. Anderson, R. and Roberts, D. (2012) 'Big Data: Strategic Risks and Opportunities: Looking Beyond the Technology Issues'. Available at: *http:// www.crowehorwath.net/uploadedFiles/CroweHorwathGlobal/tabbed_ content/Big%20Data%20Strategic%20Risks%20and%20Opportunities% 20White%20Paper_RISK13905.pdf*

4. Babcock, C. (2006) 'Data, Data, Everywhere'. *Information Week* Available at: *http://www.informationweek.com/data-data-everywhere/175801775*

5. Banerjee, S., Bolze, J. D., McNamara, J. M. and O'Reilly, K. T. (2011) 'How Big Data Can Fuel Bigger Growth'. Accenture. Available at: *http://www.accenture.com/SiteCollectionDocuments/PDF/Accenture-Outlook-How-Big-Data-can-fuel-bigger-growth-Strategy.pdf*

6. Barbaro, M. and Zeller, T. (2006) 'A Face Is Exposed for AOL Searcher No. 4417749'. Available at: *http://www.nytimes.com/2006/08/09/technology/09aol.html?pagewanted=all&_ r=2&*

7. Davenport, T. H., Eccles, R. G. and Prusak, L. (1992) 'Information Politics'. *Sloan Management Review*: 53–65.

8. Dyché, J. (2007) 'A Data Governance Manifesto: Designing and Deploying Sustainable Data Governance'. Available at: *http://www.siperian.com/documents/WP_JillDyche_DataGovernance_June07.pdf*

9. Evans, B. J. (2011) 'Much Ado About Data Ownership'. *Harvard Journal of Law & Technology* 25 (1): 70–130.

10. Grant, J. and Kirchmaier, T. (2004) 'Corporate Ownership Structure and Performance in Europe'. London School of Economics Available at: *http://eprints.lse.ac.uk/19960/1/Corporate_Ownership_Structure_and_Performance_in_Europe.pdf*

11. Halevi, G. and Moed, H. (2012) 'The Evolution of Big Data as a Research and Scientific Topic'. *Research Trends: Special Issue on Big Data* 30. Available at: *http://www.researchtrends.com/wp-content/uploads/2012/09/Research_Trends_ Issue30.pdf*

12. Hart, D. (2000) 'Data Ownership and Semiotics in Organisations, or Why They're Not Getting Their Hands on My Data'! Available at: *http://www.pacis-net.org/file/2000/377–388.pdf*

13. Holloway, S. (1988) 'Data Administration'. (Aldershot: Gower).

14. Khan, S. M. and Hamlen, K. W. (2012) 'Anonymous Cloud: A Data Ownership Privacy Provider Framework in Cloud Computing'. Available at: *http://www.utdallas.edu/~hamlen/khan12trustcom.pdf*

15. Loshin, D. (2001) 'Enterprise Knowledge Management: The Data Quality Approach'. The Morgan Kaufmann Series in Data Management Systems. Morgan Kaufmann.

16. Manyika, J., Chui, M., Brown, B., Bughin, J., Dobbs, R., Roxburgh, C. and Byers, A. H. (2011) 'Big data: The next frontier for innovation, competition, and productivity'. McKinsey Global Institute. Available at: *http://www.mckinsey.com/~/media/McKinsey/dotcom/Insights and pubs/MGI/Research/TechnologyandInnovation/BigData/MGI_big_data_full_report.ashx*

17. Marc, D., Martinez, R. and Kalaboukis, C. (2010) 'Rethinking Personal Information – Workshop Pre-read'. Invention Arts and World Economic Forum.

18. PTACT (2010) 'Data Governance and Stewardship'. e-Privacy Technical Assistance Center. Available at: *http://www2.ed.gov/policy/gen/guid/ptac/pdf/issue-brief-data- governance-and-stewardship.pdf*

19. Rajagopal, S. (2011). 'Customer Data Clustering Using Data Mining Techniques'. *International Journal of Database Management Systems* 3 (4): 1–11.

20. Robinson, N., Graux, H., Botterman, M. and Valeri, L. (2009) Review of the European Data Protection Directive.

21. SAP (2012) 'Harnessing the Power of Big Data in Real Time through In-Memory Technology and Analytics'. Available at: *http://www3.weforum. org/docs/GITR/2012/GITR_Chapter1.7_2012.pdf*

22. Schnarch, B. (2004) 'Ownership, Control, Access and Possession (OCAP) or Self-Determination Applied to Research: A Critical Analysis of Contemporary First Nations Research and Some Options for First Nations Communities'. Available at: *http://www.research.utoronto.ca/ethics/pdf/human/nonspecific/ OCAP%20principles.pdf*

23. Shields, G. (2010) 'Addressing Security and Data Ownership Issues when Choosing a SaaS Provider'. Available at: *http://www.quest.com/quest_site_ assets/ whitepapers/wpw_saas_shields_us_mj.pdf*

24. Singh, Y. J. Singh, Y. S., Gaikwad, A. and Mehrotra, S. C. (2010) 'Dynamic management of transactions in distributed real-time processing system'. *International Journal of Database Management Systems* 2 (2): 161–170.

25. WEF (2011) 'Personal Data: The Emergence of a New Asset Class'. World Economic Forum. Available at: *http://www3.weforum.org/docs/WEF_ ITTC_PersonalDataNewAsset_ Report_2011.pdf*

26. WEF (2012) 'Rethinking Personal Data: Strengthening Trust'. Available at: *http://www3.weforum.org/docs/WEF_IT_RethinkingPersonalData_ Report_2012.pdf*

27. Wigan, M. R. (1992) 'Data Ownership'. In Clarke, R. A. and Cameron, J. (eds), *Managing the Organisational Implications of Information Technology.* North Holland, Amsterdam: Elsevier.

28. Woodbury, C. (2007) 'The Importance of Data Classification and Ownership. Available at: *http://www.srcsecuresolutions.eu/pdf/Data_Classification_ Ownership.pdf*

Triggering the smart card reader supply chain

Abstract: In the last 12 years or so, many governments have launched modern identity management systems. These systems typically integrate a set of advanced and complex technologies to provide identification and authentication capabilities. The major output of such systems is smart identity cards that bind the card holders' identities to their biographical data and one or more biometric characteristics. The field of government practice has been focusing on the enrolment capabilities and infrastructure roll-out, with little focus on smart card applications in the public domain. This article attempts to address this area in the body of knowledge from a government view point. It explores card reader adoption opportunities in both the public and private sectors, and attempts to outline the United Arab Emirates' (UAE's) government's plans to disseminate card readers and promote their adoption in government and various industrial groups in the country.

Keywords: *Identity management; electronic identity; eID; smart cards; card readers.*

1. Introduction

'There were 30 billion plastic cards issued in 2011 alone. If you were to line up the cards end-to-end, that amounts to nearly 1.6 million miles of plastic every year – enough to create a six-lane plastic highway between the Earth and the Moon. Mobile wallets in the long run have the potential to replace billions of plastic cards that end up in landfills, while providing a secure and convenient user experience'.

The public and business sectors are showing greater interest in plastic card adoption. However, there is a shift towards a more sophisticated plastic card in the industry, namely smart plastic cards (2, 3, 4). Smart cards provide more advanced capabilities in identification, authentication, data storage and application processing (5-8). Smart cards contain electronic memory and, in some cases, an embedded integrated circuit. They basically come in two types; contact and contactless. Contactless smart cards only require close proximity to a reader to enable the transmission of identification data, commands and card status, hypothetically taking less time to authenticate compared to contact smart cards that require insertion into a smart card reader (9). Each reader and card has an internal antenna that securely communicates with the other.

Smart card authentication is considered to be the best-known example of a proof by possession mechanism (10, 11). Other traditional classes of authentication mechanisms include proof by knowledge (e.g., passwords) and proof by property (e.g., fingerprints). When combined and used for user authentication, smart cards can help improve the security of a device as well as provide additional security services (12-14). Figure 8.1 depicts a typical user logon and authentication process using a smart card.

A considerable number of governments around the world have launched modern identity management programs in an attempt to create a more accurate population register, and to develop identification and authentication capabilities (15). Governments have been promoting their national identity schemes somewhat aggressively to gain public acceptance

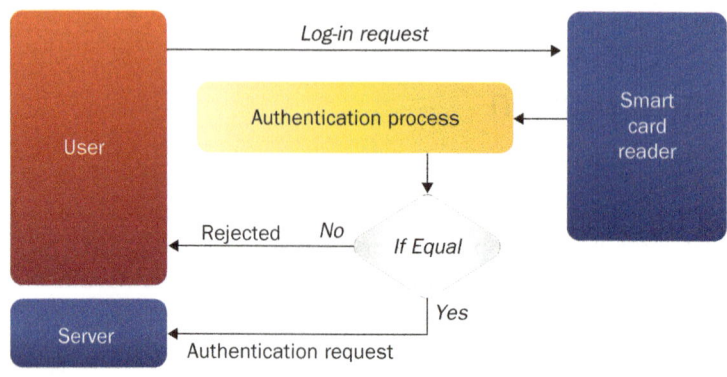

Figure 8.1　User authentication with smart cards

and secure necessary funding. Among the numerous benefits reported by governments is that the new smart identity cards may replace many of the existing identification cards, such as driving licences, health insurance cards, and so on (16-18), and therefore have a high impact on lowering costs and minimising duplication in widely produced identity documents. Smart cards are also promoted on the basis of having the potential to become an innovative approach in addressing the e-government and e-commerce needs of strong authentication.

In reality and practice, however, governments have been paying significant attention to population enrolment and infrastructure roll-out with little attention to widening the smart identity card applications in the public domain. It is interesting to refer here to the fact that it is estimated that governments around the world have issued hundreds of millions of smart cards in the last ten years in the form of national identity cards. Despite this gigantic number of smart cards issued by governments, the field of practice does not show any noteworthy government action that can truly justify such massive spending and show clear economic value.

On the other hand, market research shows that governments will still spend increasing and capacious financial budgets on smart card acquisition in the coming years (19-21). A recent study by Frost and Sullivan argues that the field of smart cards and their adoption will remain strong and may exhibit renewed potential for all markets, given the growing concern for electronic security, and increasing overall system performance with contactless technology (22). The study also anticipated that all of the four primary markets which smart card readers address (i.e., government, banking and payment, transportation, and corporate security) will continue to grow. Our experience in the field strongly supports this finding. We envisage that smart card readers have huge market penetration and adoption opportunities, provided that the right strategies and business cases are put in place. In this article, we attempt to outline the UAE's plans to roll out smart card readers and promote their adoption in the public and private sectors.

This article is structured as follows. Section 2 provides a short overview of the UAE's identity management infrastructure and the features of the smart identity card issued as part of the UAE's scheme. Section 3 outlines how the demand for card reading by service providers may impact the supply of card readers. Section 4 provides an overview of the role of the UAE's identity issuing authority in the card reader market. Finally, Section 5 concludes the article with some discussion.

2. The UAE: a regional leader in electronic identity (e-ID)

The UAE is considered to be a regional leader in the field of digital identity management infrastructure (23). Through its national identity issuing authority set up in 2004, to date it has issued more than 8.5 million digital identities to its citizens and resident population. Digital identity is linked to the individual's biographical data and his/her facial and fingerprint biometrics. The digital identity profile is represented by a set of digital certificates and strong authentication credentials, packaged in the form of a secure smart identity card.

The UAE identity card is essentially a Java-based smart card packed with features such as contact and contactless, multi-factor authentication and a challenge response mechanism for determining the card holder's identity and whether or not the card is genuine. Built on open standards, the Java-based operating system card is secured by asymmetric keys and specific protocols for communication. It uses both match-on-card and match-off-card features to provide various authentication capabilities.

In six years, the UAE has achieved the target of enrolling almost 98 per cent of its citizens and the resident population and issuing them with smart digital identity cards. The new smart identity card is currently a pre-requisite in order to access any services in the government and public sectors, and, in part, in the private sector as well.

The government has distributed around 10,000 smart card readers in the last five years to the public sector and some private sector organisations to allow them to electronically 'read' the data stored on the smart identity card. Early adopters of smart card readers reported that they have succeeded in shortening over-the-counter customer transactions from 10 to 20 minutes on average (24, 25).

It is envisaged that smart card readers and applications will be further cultivated in the UAE in the next few years and will become an integral part of enterprise architecture in both the public and private sectors due to their potential to improve service delivery systems and secure access to networks and online transactions. The simple 'read' need of the smart card is also expected to evolve with the need to undertake stronger identification, verification and authentication procedures to confirm the identity of the service seeker. We explore this further in the next section.

3. ID card readers and the reader supply chain in the UAE

As indicated earlier, modern identity card programs have been launched by many governments with a strong belief that they will address many of the challenges and national development priorities (15). However, no country has linked the use of smart card readers with identity verification needs as an imperative for delivering government and public sector services. Besides, some countries issue digital certificates on their identity cards but, since they are exportable, the card itself becomes redundant in certificate usage.

The government of the UAE has been working to develop secure authentication capabilities to support e-government transformation and e-commerce initiatives. This transformation is sought as part of the UAE's e-government strategy 2012–2014 which aims to develop innovative delivery systems and support the development of a digital economy (26, 27). The UAE's new smart identity card is designed to support this particular objective of authentication as it contains the digital credentials of the card holder for use both in in-person and online environments.

Currently, the UAE mandates the physical 'presence' of the new smart identity card to benefit from any service that is provided by government and public sector agencies. With increasing interest in reading data electronically from the chip, card reading capabilities are becoming a necessity, creating more demand for card readers. A quick look at smart card reader availability around the world seems to reveal that they are not commodity items that are available in electronics stores. So, one may wonder, why is it that one does not see the readers in retail stores? Why are readers not available like mobile phones? The reason is obvious when the supply chain for identity card readers is analysed. The use of the card reader is a direct function of the need to read the card data. Table 8.1 attempts to illustrate this.

As is evident from the table, the predominant user of the card reader is the service provider and the card reader ownership very much depends on the service delivery channels. Adoption of identity card readers by service providers thus becomes the key to the widespread availability of card readers. Figure 8.2 below depicts the ID card reader supply chain as seen in the UAE. Rows two, three and four of the supply chain diagram look like the conventional chains of the manufacturer, the supplier (distributor) and the retailer.

Table 8.1 Application types and card reading needs

Application	Channel	Need Card Reader	Reader Owner
Simple identification – no need for data entry: manual check of the ID Card and manual reading of IDN.	Physical	NO	None
ID required to be entered as data.	OTC Kiosk Web portal	YES YES YES	Service provider Service provider Card Holder
Service to be delivered OTC – ensure that it is being delivered to the correct person.	OTC	YES	Service provider
Service to be delivered OTC – ensure that it is being delivered to the correct person and require confirmation of service delivery (signature of service beneficiary).	OTC	YES	Service provider
Service requested remotely.	Kiosk Web portal	YES	Card Holder
Service to be delivered remotely-ensure it is being delivered to the correct person and require confirmation of service delivery.	Kiosk Web portal	YES	Card Holder Service provider

In general terms, the ID card reader supply chain is seen here to be driven by push and pull forces. The push and pull represents the need for identity services (i.e., verification, validation and authentication needs) to enable the delivery of services by various service providers. The electronic service channels dictate the integration of smart card readers that facilitate the use of the government issued smart ID card. Smart card adoption in retail and distribution is expected to further support growth opportunities in this critical industry.

Retailers and distributors in the UAE have placed increased importance on conducting business in a more cost effective and time efficient manner. Intensifying competition, cost/scarcity of retail space and growing

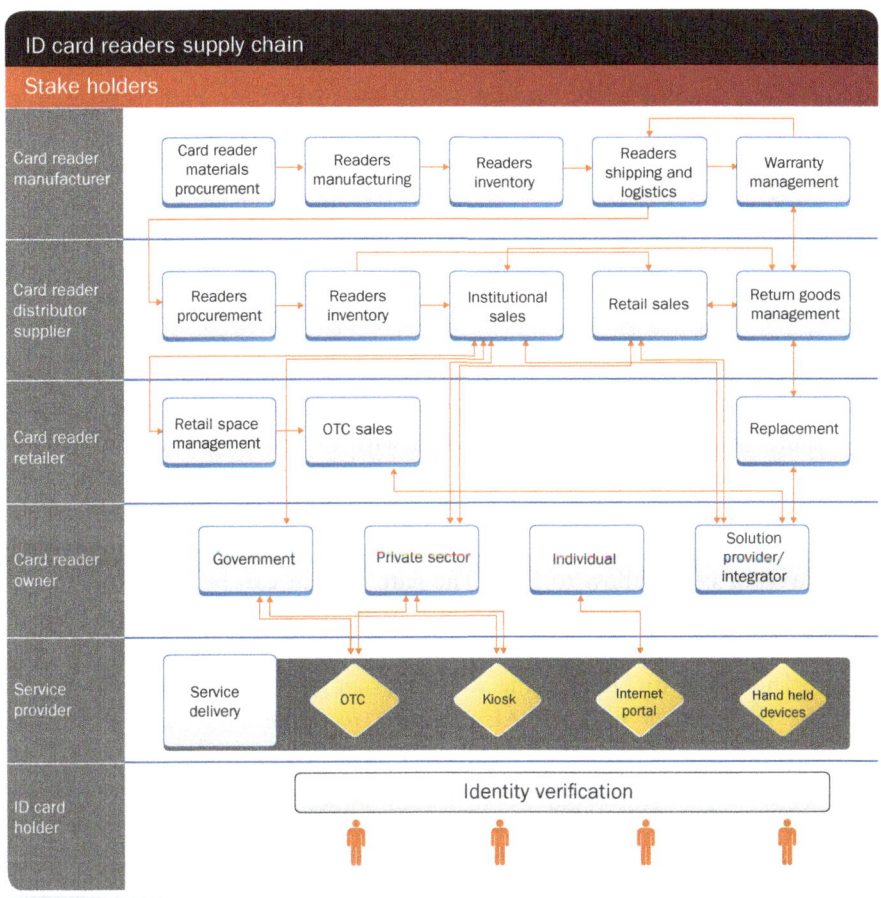

Figure 8.2 Identity card reader supply chain

customer expectations are driving retail and distribution businesses to establish collaborative business models – necessitating the formation and maintenance of new relationships. It is in this context that we can view the efforts of the UAE to popularise the use of the smart ID card and drive the ID card reader supply chain. The next section elaborates on this further.

4. The role of the UAE Identity Authority in the ID card reader market

Although UAE smart identity cards have been made mandatory to access all government services, there is no clear and perceptible adoption of smart card readers among the service providers. There is much more reliance on the physical 'by eye look' rather than on the 'electronic reading and processing' capabilities of the smart card. Here is where the UAE government, through its Identity Authority,* is currently working to play a significant role in promoting and supporting card reader adoption in both the public and private sectors. The authority's role and current activity is akin to the trigger mechanism used for thyristors to become conductors of current. See Figure 8.3.

A thyristor is a semi-conductor device with special characteristics. It is the equivalent of a mechanical switch and consists of two diodes. The special characteristic of a thyristor is that it acts as a non-conductor of current till it has such a voltage across its anode and cathode that the semi-conductor itself breaks down or with a small trigger current applied at its 'gate' allows a large amount of current to be conducted almost instantaneously. This trigger current is called the gate current and is usually in milliamperes, compared to the thousands of amperes that the thyristor could allow to flow. The gate current can be withdrawn once the conduction starts and it does not switch the thyristor from conducting to non-conducting.

The similarity between the above current gate concept and that which the government of the UAE is trying to achieve is remarkable. We envisage the UAE Identity Authority to play a triggering role and, as mentioned above, in order to get the thyristor to turn on we need to inject a pulse of current into the gate. Much as a gate current (a very small amount) is required for the conduction of large current flows, a trigger is needed to set in flow identity card-based services and the adoption of card readers.

Government regulations may be seen here to act as having the potential to trigger a higher demand for smart card readers and the widespread

---

*The Emirates Identity Authority is a federal government entity in the UAE established in 2004 with the task of developing a modern national identity management infrastructure. The first phase in the scheme aims to enrol the whole population, around 8.5 million, and issue them with biometric smart identity cards. Having completed the enrolment phase, the authority is currently working on maximising the reach and adoption of its identity cards both in the public and private sectors. The argument is that smart cards have significant potential to contribute towards creating value-added customer services and operational efficiency, while providing advanced authentication capabilities.

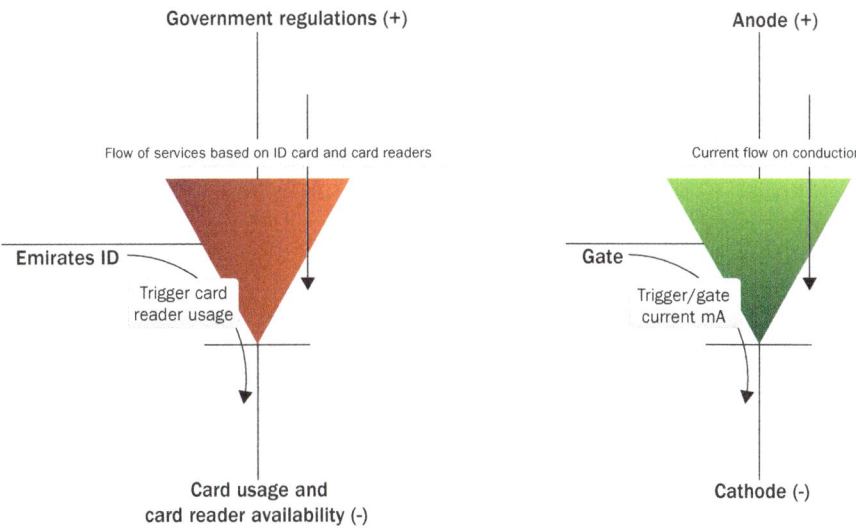

Figure 8.3   Card readers and the gate current concept

application and usage of the identity card and the card readers in various industries. Nonetheless, inertia (e.g., lack of awareness, lack of applications and so on) will act as resisting elements to the roll-out of services.

The UAE Identity Authority recently announced that it will distribute tens of thousands of card readers across different government service providers. It is considered that this will set in motion a supply chain movement. In other words, once a certain number of readers are available in the public domain and used as a requirement to access government services, the public and financial sectors are expected to follow suit. As identity card usage becomes more prevalent, it is expected to cascade to a wider availability of services on web portals as e-services.

While the service providers themselves would act as institutional buyers in the supply chain, the vendors of the card readers are expected to become more visible in the retail industry. This would bring the retail buyer, that is, the individual buyer, to create the demand to procure the readers.

The healthcare sector is another subdivision that would gain immensely from the availability of identity card readers to enable identity card-based healthcare transactions. The smaller players in the healthcare supply chain, like pharmacies, clinics and so forth, would act as individual buyers and would drive the retail market towards identity card readers. Banks and other financial institutions would be major beneficiaries from fraud containment, while citizens and residents would be considered as direct beneficiaries of the convenience of economic transactions involving identity verification. This is expected to result in the economic activity of millions of card readers in the country. At an average retail price of $13.50 per reader, this would translate into tens of millions of dollars just in terms of hardware sales.

Indeed, this is considered to be a new segment of economic activity that is currently absent. The UAE Identity Authority's action of buying a few thousand readers and placing them with government service providers would trigger this economic activity in the space of the next two years. This is just a part of the whole. Along with the readers would come the services directly related to the hardware in terms of warranty and technical services. The software and application development of integrating the identity card and the card reader into applications is a supplementary economic activity. At a conservative estimate, this activity could be worth tens of millions of dollars. In total, the UAE government's gate trigger role would result in an economic activity worth hundreds of millions of dollars and would contribute directly to the overall GDP of the country. Besides, card readers are foreseen to be available in supermarkets and electronics stores as well. The digital literacy of the card holders would be a direct result of using identity cards.

On the other hand, the impact of a wider adoption of smart card readers in various industries in the UAE may well trigger more adoption in the Gulf Co-operation Council (GCC) countries and the region as a whole. Improving customer service and internal efficiency will be the key drivers for this adoption. The hi-tech and security features of the smart identity card would result in a secure environment which would contribute to containing fraud and identity theft. The economic activity stemming from smart card distribution in this case will then become a factor in the calculation. Identity card reader manufacturers, distributors and electronics retailers would well be advised to gear up to this reality. The early players would have the benefit of being early movers in exploiting this niche market.

5. Conclusion

Smart cards are likely to become indispensable tools in the future in addressing the competitive interests of diverse groups of industries. Government issued identity cards are envisaged to play a critical role in this area. Their advanced authentication capabilities are likely to promote higher chances of acceptance by both service providers and the public. However, governments will need to put in place clear strategies as to how they intend to support the identification and authentication requirements both in the public and private sectors.

In this article, we have attempted to outline and bring about some insights from a government's field of practice. The UAE government's plans should provide some understanding of how a government views its role in promoting the use and application of its smart identity cards.

We envisage that governments, specifically those implementing modern identity card systems, will show an increasing interest in card readers in the next few years. Many governments will work towards replacing existing identification documents with the new smart identity card. As smart card readers naturally demonstrate stronger authentication capabilities, this in itself will support higher levels of trust and participation. This may also have a significant impact on the progress and development of e-government and e-commerce business models. To a great extent, identity theft and transaction fraud can be controlled too.

Besides, we also believe that governments will attempt to regulate the computer industries in their countries to integrate personal computers with smart card readers. This is likely to be seen either as part of a computer keyboard or driver or perhaps as something external (28). Smart cards have the potential to reshape service delivery and the way in which services are provided, both in the government and public sectors as well as in the retail and distribution industries. The rapid technological pace in the smart identity card industry will not only revolutionise the future of identification and authentication but will also open up new business opportunities and create new economic niches.

References

1. Narendra, S. (2012) 'Connecting identity and mobility: A secure, scalable and sustainable mobile wallet approach'. *IQT Quarterly* 4 (1): 18–21. Available at: *http://tyfone.com/IQT_Quarterly_Summer2012_Tyfone_article.pdf*
2. Hansmann, U. and Nicklous, M. S. (2002) 'Smart Card Application Development Using Java'. Berlin: Springer.
3. Morgner, F., Oepen, D., Müller, W. and Redlich, J. (2012) 'Mobile smart card reader using NFC-enabled smart phones'. Available at: *http://sar.informatik.hu-berlin.de/research/publications/SAR-PR-2012-07/SAR-PR-2012-07.pdf*

4. Paret, D. (2005) 'RFID and Contactless Smart Card Applications'. New York: Wiley.

5. Mayes, K. and Markantonakis, K. (eds) (2010) 'Smart Cards, Tokens, Security and Applications'. Heidelberg: Springer.

6. Pelletier, M., Trepanier, M., and Morency, C. (2012) 'Smart Card Data in Public Transit Planning: A Review'. Available at: *https://www.cirrelt.ca/ DocumentsTravail/CIRRELT-2009-46.pdf*

7. Rizvi, S. A. M., Rizvi, H. S. and Al-Baghdadi, Z. (2010) 'Smart Cards: The Future Gate'. *Proceedings of the World Congress on Engineering and Computer Science*: 1. WCECS, October 20–22, San Francisco, USA. Available at: *http://www.iaeng.org/publication/WCECS2010/WCECS2010_ pp81–86.pdf*

8. Taherdoost, H., Sahibuddin, S. and Jalaliyoon, N. (2011) 'Smart card security; Technology and adoption'. *International Journal of Security (IJS)* 5 (2): 74–84. Available at: *http://cscjournals.org/csc/manuscript/Journals/ IJS/volume5/Issue2/IJS-84.pdf*

9. Finkenzeller, K. (2010) 'RFID Handbook: Fundamentals and Applications in Contactless Smart Cards, Radio Frequency Identification and Near-Field Communication'. New York: Wiley.

10. Jansen, W., Gavrila, S., Séveillac, C. and Korolev, V. (2005) 'Smart Cards and Mobile Device Authentication: An Overview and Implementation'. National Institute of Standards and Technology Available at: *http://csrc.nist. gov/publications/nistir/nist-IR-7206.pdf*

11. Ramasamy, R. and Muniyandi, A. P. (2012) 'An efficient pass-word authentication scheme for smart card'. *International Journal of Network Security* 14 (3): 180–186. Available at: *http://ijns.femto.com.tw/contents/ ijns-v14-n3/ ijns-2012-v14-n3-p180-186.pdf*

12 Baboo, S. S. and Gokulraj, K. (2010) 'A secure dynamic authentication scheme for smart card based networks'. *International Journal of Computer Applications* 11 (8): 5–12. Available at: *http://www.ijcaonline.org/volume11/ number8/pxc3872157.pdf*

13. Bakker, B. (1999) 'Mutual authentication with smart cards'. Available at: *http://static.usenix.org/events/smartcard99/full_papers/bakker/bakker.pdf*

14. Monk, J. T. and Dreifus, H. N. (1997) 'Smart Cards: A Guide to Building and Managing Smart Card Applications'. New York: Wiley.

15. Al-Khouri, A. M. (2012) 'Population growth and government modernisation efforts: The case of GCC countries'. *International Journal of Research in Management and Technology* 2 (1): 1–8.

16. Briggs, J. and Beresford, R. (2009) 'Smart cards in health'. Available at: *http://www.chmi.port.ac.uk/pubs/smartcards/smartcard_report.pdf*

17. Chellappan, S. and Paruchuri, V. (2009) 'Integrating Smart Cards with Vehicular Networks: Architecture and Applications'. Available at: *http:// web.mst.edu/~cswebdb/Workshop-AFRL/paper22009751.pdf*

18. Choudary, O. S. (2010) 'The Smart Card Detective: a hand-held EMV interceptor' Master's Dissertation University of Cambridge. Available at: *http://www.cl.cam.ac.uk/~osc22/docs/mphil_acs_osc22.pdf*

19. Catherine, A. and Barr, W. J. (eds) (1996) 'Smart Cards: Seizing Strategic Business Opportunities'. McGraw-Hill.

20. Hendry, M. (2007) 'Multi-application Smart Cards: Technology and Applications'. Cambridge: Cambridge University Press.
21. Smart Card Alliance (2011) 'Smart Cards and Biometrics'. Available at: *http://irisid.com/download/news/Smart_Cards_and_ Biometrics_030111.pdf*
22. Frost and Sullivan (2010) 'World smart card readers and chipsets market'. Available at: *http://www.frost.com*
23. WRA (2012) 'Largest biometric database: Emirates Identity Authority sets world record'. World Record Academy Available at: *http://www. worldrecordacademy.com/technology/largest_biometric_database_ Emirates_ Identity_Authority_sets_world_record_113117.html*
24. Al-Bayan (2009) 'Identity card shortens the time of data entry in the Dubai Courts to 7 seconds'. Al Bayan Newspaper, Dubai, UAE (Arabic Version). Available at: *http://www.albayan.ae/across-the-uae/1235654452291-2009-03-02-1.410704*
25. Al-Khouri, A. M. (2012) 'Identity and mobility in a digital world'. *Investment & Technology* 4 (1).
26. Al-Khouri, A. M. (2012) 'e-Government strategies: The case of the United Arab Emirates (UAE)'. *European Journal of e-Practice* 17: 126–150.
27. Al-Khouri, A. M. (2012) 'Emerging markets and digital economy: Building trust in the virtual world'. *International Journal of Innovation in the Digital Economy* 3 (2): 57–69.
28. Mohammed, L. A., Ramli, A., Prakash, V. and Daud, M. D. (2001) 'Smart card technology: Past, present, and future'. Available at: *http://www.journal. au.edu/ijcim/2004/jan04/jicimvol12n1_article2.pdf*

9 781909 287631